AF681720

Quality and risk management in agri-food chains

Quality and risk management in agri-food chains

edited by:

Brigitte Petersen

Manfred Nüssel

Martin Hamer

Wageningen Academic Publishers

EAN: 9789086862368
e-EAN: 9789086867899
ISBN: 978-90-8686-236-8
e-ISBN: 978-90-8686-789-9
DOI: 10.3920/978-90-8686-789-9

First published, 2014

© Wageningen Academic Publishers
The Netherlands, 2014

This work is subject to copyright. All rights are reserved, whether the whole or part of the material is concerned. Nothing from this publication may be translated, reproduced, stored in a computerised system or published in any form or in any manner, including electronic, mechanical, reprographic or photographic, without prior written permission from the publisher:
Wageningen Academic Publishers
P.O. Box 220
6700 AE Wageningen
The Netherlands
www.WageningenAcademic.com
copyright@WageningenAcademic.com

The individual contributions in this publication and any liabilities arising from them remain the responsibility of the authors.

The publisher is not responsible for possible damages, which could be a result of content derived from this publication.

Table of contents

List of figures

List of tables

Preface

Effective quality and risk management is a topical issue that academia and industry is equally concerned with at the moment. These two management tasks belong together. Many companies in the agri-food industry but also public institutions have become painfully aware of the negative consequences of insufficient quality and risk management at the latest since the dioxin and EHEC crisis in 2011. It will be a challenge for the future to avoid negative impacts on all parts of the food chain starting with the environment as the main resource of food production and ending with the consumer.

Crises have quite often been triggered by avoidable management errors resulting from a not yet sufficiently developed quality and risk culture, on the one hand, and from a still existing attitude on the part of the managers of companies that quality and risk management are only annoying duties imposed either by legislation or by customers.

The objective of an economically oriented quality and risk management therefore must be to create transparency in the business and production processes and thus enable the management to recognise changes of risks and chances in good time and derive adequate options for action enabling quality and risk control. In addition to this, an integrated quality, environmental and risk management system forms the basis for fulfilling the increasing number of legal requirements, industry-specific standards and customer specifications as well as consumer demands.

Quality management (QM) is a very complex topic, however, which requires just as much knowledge about the statutory framework and numerous standards as well as knowledge about the fields of application for QM methods and an understanding of complex and intercompany production and value models. Qualitative aspects such as the importance of a coordinated quality policy at all stages of the value chain – from agricultural primary production and food processing to food retail – and the design of a common risk culture also play an important role in this regard. In addition to this, decision-theoretical foundations for structured handling of crisis situations are indispensable.

This book follows on the necessities and challenges mentioned by providing an up-to-date overview of queries, approaches and problem solutions in quality and risk management of individual enterprises and across enterprises in the agri-food industry. This book addresses both, representatives from companies and from public and private organisations as well as teaching staff and students.

We would like to take this opportunity to express our gratitude to all co-authors for their vigorous and competent support and to Mike Jacobs from Wageningen Academic Publishers for his dedication and personal commitment as critical discussion partner and motivator. Our most sincere thanks goes to the doctoral candidates of the Universities of Bonn, Bayreuth, Göttingen and Wageningen who have dealt with various aspects of quality and crisis management in recent years within the framework of their doctorate since many results and important findings from their work have been incorporated into the texts and illustrations of this book.

Berlin – Bonn – Wageningen, January 2014
Brigitte Petersen, Manfred Nüssel, Martin Hamer

Chapter 1. Introduction

B. Petersen[1], M. Nüssel[2] and M. Hamer[3]
[1]Universityof Bonn, Katzenburgweg 7-9, 53115 Bonn, Germany
[2]DRV, Deutscher Raiffeisenverband e.V., Pariser Platz 3, 10117 Berlin, Germany
[3]GIQSe.V., c/oUniversityofBonn, Katzenburgweg 7-9, 53115Bonn,Germany
b-petersen@uni-bonn.de

In the agri-food industry, quality management (QM) is meanwhile regarded as a cross-sectoral task. Interdisciplinary research is undertaken in science in this respect, and together with enterprises, solutions are developed for the industry. Academia and industry alike are pursuing a holistic quality strategy demanding a far reaching change in the attitude and behaviour of everybody involved in addition to providing the answers to economic, organisational, technical and scientific queries.

This book answers the question why coordinated quality management and communication principles between individual enterprises but also between industry, academia and public authorities are necessary for this above all in today's globalised world. The authors of this book explain how QM systems tailored to the industry as a basis for efficient value processes and also as the basis for international trade with feed and foodstuff may look. Furthermore, they present in detail innovations for designing quality communication in value chains. These innovations were developed in recent years in interdisciplinary research networks to strengthen the relationships between customers and suppliers to quickly tackle crisis situations jointly and to withstand the global competitive pressure with renewed strength.

In addition to this, the authors illustrate thoroughly the numerous reasons, why measures – in terms of preventive quality assurance – are increasingly transferred to the planning preliminary product and process domains.

In this respect, the importance of innovation alliances, strategic partnerships and regional clusters in the field of quality communication is analysed in detail. These research alliances line up to promote information and communication technologies, technical, organisational and legal queries and the development of test methods and procedures in the area of quality sciences. Global consensus building in connection with standardisation issues and the

B. Petersen, M. Nüssel and M. Hamer (eds.) *Quality and risk management in agri-food chains*
DOI 10.3920/978-90-8686-789-9_1, © Wageningen Academic Publishers 2014

elaboration of an industry-specific roadmap to technically support time-critical processes in quality and crisis management is also a topic. Innovative technical and organisational solutions are indispensable when it comes to testing feed and foodstuff rapidly and during the production process, communicating results quickly, taking decisions immediately and finally selecting the right alternatives for action. This not only applies to commercial enterprises and supply chains to food retail but also to public authorities in the area of food monitoring and veterinary authorities in the field of disease prevention and control.

In several chapters, the authors of this book broach the issue of entrepreneurial activities always being associated with risks, and this applies to the agri-food industry as well because the consequences of many management decisions are uncertain and not all market developments or conditions under which production may take place can be completely predicted. Risk management is considered in this respect as part of company management and thus as a management duty within QM.

Risks can be seen as possible deviations from planned corporate objectives. Even though this general definition also includes positive deviations from the objectives, i.e. possibilities, negative deviations and thus the risks and threats are in the foreground here when having a closer look at risk management. Risks in this sense can encompass the whole spectrum from petty offenses to existential crises. Above all, natural catastrophes, lost harvests, animal diseases and other risks can cause great damage potential which can threaten the existence of complete sectors such as the meat, milk, fruit and vegetable and feed industry. Many interrelations within international value chains make assessment of the potential extent of damage caused by a crisis more difficult.

However, crises for individual companies and for individual sectors also arise as a result of quality problems, uncertainty among customers, turnover and yield losses and due to considerable cost for image campaigns to win back consumers. In the following chapters it will be explained why risks will intensify and escalate into company crises if they are not recognised and controlled in time. Crises thus result from losses occurred and describe an undesired current situation of an individual company or a network of companies while risks present future hazard potentials.

The authors of Chapters 2 and 4 exhaustively explain the most important objectives and motivation reasons underlying the introduction of quality and risk management systems (e.g. long-term and sustainable customer loyalty and safeguarding the existence, providing room for manoeuver, securing the planned corporate objectives and reducing risk and capital cost). When doing so, first of all in Chapter 2, the central terms, objectives and duties of QM are clarified and the most important developments in QM of the past decades are traced. In addition to this, in Chapter 3 the statutory framework is presented. It is illustrated which legal requirements apply to enterprises in the agri-food industry if they produce or trade foodstuff. Finally, Chapter 4 presents the models used most frequently to clarify interdependencies and strategies of QM deduced from quality sciences.

Chapter 5 deals with those standards used in recent years to establish QM systems in various organisations of the industry and check them for conformity. At the same time, this chapter addresses the issue of historical development of standardisation activities in this area and refers to trends to be expected in future.

To meet the Millennium Development Goal (MDG) on reducing hunger and other environmental objectives, global food production will have to increase, food distribution is to be improved and agriculture needs to reduce its current environmental impacts (UNEP 2012). In connection with these goals, several aspects regarding food chain management issues and systems will be described in Chapter 6. Information and communication systems in food chains as well as traceability, innovation management in agro-food networks, cold chain management and environmental management in food chains will also be discussed in this chapter.

In line with the One Health Concept, Chapter 7 will provide an overview and initiates scientific discussions focused on novel analytical and organizational approaches in heard health management and in food quality and safety, food authenticity and food integrity. Strategies enabling to address current and emerging issues will be introduced. One Health is presented as an integrated approach to health concentrating on the interactions between animals, humans and their various environments. It encourages collaborations, synergies and cross-fertilisation of all professional sectors and actors in general whose activities may have an impact on healthy food and water, public health and health services.

The focus of the book is on concrete case studies. By means of flowcharts it is described which instruments and procedures support implementation of quality and risk management systems in practice. The relevant Chapters 8 and 9 contain experiences from scientists and those responsible from industry which have jointly designed, introduced and continuously improved intercompany quality management systems in many collaborative projects. Particular attention is paid to sector-specific organisational structures. In the course of this, the role of network coordinators will also be investigated thoroughly since they already undertake intercompany management tasks such as coordination of supplier audit and crisis management within the supply chains to food retail. The authors make recommendations and examine issues within the framework of projects such as the following:

- Which repeatedly occurring, inhibiting and supporting factors have to be taken into account during implementation?
- Who is to acquire which competences when taking over tasks and responsibility in quality and risk management?
- Who assumes responsibility for core processes in intercompany QM systems?

Furthermore, these two chapters will offer interpretation aids as to how international sector-neutral standards can be translated into the language and designations of the agri-food industry in German-speaking countries.

The book concludes with a perspective of future challenges and chances. It is predicted that also in the future dynamics and adaptability will be the most important success criteria for applied and efficient quality management systems. In global business relations of companies in the agri-food industry, it will remain a primary objective to establish quality and sustainability as key factors. If it is possible, on top of this, to actively promote the idea of quality by training talented managerial and junior staff in quality and crisis management, the global competitiveness of the industry will also be strengthened.

Chapter 2. Key terms and major developments

Quality is an increasingly important competitive factor in the agri-food industry. For a critical debate, it is necessary to define the terms used in this context and to highlight the developments of quality management achieved in recent decades.

This chapter will inform you about:
- the definitions of the most important QM terms;
- which structures have become established in the food supply chains; and
- how and why quality strategies in the agri-food industry have changed and will change.

Keywords: quality strategies, definition of quality, quality standards, quality management systems, value network, development of quality

Summary

The intercompany terminologies of quality management have been interpreted by companies of the agri-food industry in the past few years referred to their sector-specific production and business processes and to the customer-supplier relationship. It is characteristic for the production of foodstuffs that it is performed in several interlinked production stages separated geographically and organisationally from each other. Currently, dynamical intercompany quality networks develop to respond quickly and targeted to social and global economic changes on global markets. This reflects a change in quality management from quality control to dynamic quality management networks including the approach of Total Quality Management. Team-oriented quality management methods such as HACCP, FMEA and Six Sigma are gaining special importance. The EU Food Law allows for self-monitoring systems according to the HACCP concept. Different standards for quality communication between the individual stages of the value chains have been developed on a voluntary basis and on the initiative of industry.

DOI 10.3920/978-90-8686-789-9_2, © Wageningen Academic Publishers 2014

2.1 Definition of quality and its significance

V. Schütz[1], S. Lehnert[2] and M. Nüssel[1]
[1]DRV, Deutscher Raiffeisenverband e.V., Pariser Platz 3, 10117 Berlin, Germany
[2]University of Bonn, Katzenburgweg 7-9, 53115 Bonn, Germany
schuetz@drv.raiffeisen.de

The importance of quality in the agri-food industry has grown steadily in the past few years. The reasons for this are increasing market saturation and growing competitive pressure caused mainly by the rising number of procurement and sales markets along with globalisation of the flow of goods. Therefore, quality is one of the decisive differentiating features in global customer-supplier relationships.

The term, *quality*, however, and thus the understanding of the concept of quality may have a different meaning and hence significance for individual users or consumers of a product. For the terms derived from the Latin word qualitas (= texture, characterstic, feature, status), there exist many definitions, above all since in the 1950s, quality assurance was established in the automotive and aerospace industry. Nowadays, in the food supply chains quality is defined in DIN EN ISO 9000:2005 as 'degree to which a set of inherent characteristics fulfils all requirements'. The associated terms are also defined in the DIN standard (DIN EN ISO 9000, 2005) and are subsequently derived from it.

The *requirements* demanded from a process or a product regarding its stated, obligatory or generally implied quality are manifold and are determined by three different perspectives: by the customer's point of view, the organisation's internal point of view and the point of view of further stakeholders (e.g. insurance companies, banks, public institutions, non-governmental organisations (NGOs)).

A *characteristic* 'is a distinguishing feature' of a product (DIN EN ISO 9000, 2005). This feature can be inherent to the product and is thus objectively measurable, e.g. weight, size or density. In contrast to this, the 'characteristic assigned to a product or process or system (e.g. the price of a product, the owner of a product) is not a quality characteristic of that product or process' (DIN EN ISO 9000, 2005). The DIN EN ISO 9000:2005 standard differentiates between six more classes of characteristics: physical (chemical, biological), sensory, temporal, behavioural, ergonomic and functional. The characteristics of food and primary products are found mainly in the three classes mentioned first.

- physical, i.e. mechanical electronic features (e.g. conductivity of milk);
- chemical (e.g. trace elements in the product);
- biological (e.g. number of bacteria on the surface);
- sensory (e.g. flavour of wine, cracking noise when eating wienerwurst);

- temporal with regard to availability, time performance, dependability, etc. (e.g. delivery time of just-in-time deliveries, period until goods can be delivered).

For the area of personal services, the characteristics class *behavioural* (e.g. courtesy, honesty, veracity) is important since these are closely related with personal appearance. These quality characteristics of a company play an important role for food production in particular during times of crisis. Confidence of customers in the food industry and food retail is crucial to enter new markets, to win them back in times of crisis or to maintain them (see Chapter 9).

A *product* is the result of a process (see Chapter 4). A differentiation is made between material and immaterial products. Consulting services are for example typical immaterial products. Material products may be raw materials, semi-finished and finished products or by-products of food production. Until a foodstuff will reach the end consumer at the very end of an often long-lasting value process, a combination of processes has to be designed which result in different raw materials, semi-finished and finished products and services. In contrast to the distinction of products made in business management, the DIN standard (DIN EN ISO 9000, 2005) differentiates between four product categories: services, software, hardware and processed materials. The category 'hardware' encompasses all inputs necessary for the process. Wheat would be for example the necessary hardware to grind flour. Auxiliary materials and consumables essential for the process are according to the classification of 'processed materials' products, e.g. lubricants. The category of software includes for example computer programs or dictionaries.

A *process* is defined as a 'set of interrelated or interacting activities which transforms inputs into outputs' (DIN EN ISO 9000, 2005). Inputs in a process are usually products of previous processes (see Chapter 4.1). This means that not only a single process has to be planned to achieve a certain result, but an adjustment to previous processes has to be carried out as well to accomplish the expected added value. Objective requirements regarding the expectations of the characteristics are necessary to audit the results of a process. These are defined in so-called specifications in quality management. *Specifications* are therefore documents stating the requirements on a product, a process or a system. This necessitates, on the one hand, that characteristics are defined and quantified by determining lower and upper limits, thus identifying a reference value or reference range. On the other hand, an applicable audit methodology has been selected.

DIN EN ISO 9000:2005 neutrally refers to an *organisation* in connection with quality management. This means 'a group of people and facilities with an arrangement of responsibilities, authorities and relationships' (DIN EN ISO 9000, 2005). Normally, an organisation is structured, and can be of a public or private nature. Legally, these are, for example, societies, corporate bodies, enterprises, companies, institutes, non-profit organisations or associations. Furthermore, an organisation can consist of parts or hybrids of such institutions. The quality manual describes the relevant organisation in detail, and thus a scope of application is determined for the quality management system. The *organisational structure* reflects 'the arrangement of responsibilities, authorities and relationships between people' (DIN EN ISO 9000, 2005). The organisational structure presents the interfaces between inter-organisational hierarchies and inter-organisational interrelations. Furthermore, the organisational structure and areas of responsibility can be derived from this. The top management includes those people managing and controlling a company. Within the framework of quality management (Chapter 2.2), this group plays a key role. It is responsible for quality policies, quality planning, quality objectives, quality assurance and quality improvement.

The *quality policy* of an organisation provides the framework and determines the overall objectives in terms of a quality strategy. The focal points of quality policies are the requirements and expectations of customers, and, however, not the material, personal and financial resources needed for its implementation. These will be considered in *quality planning* converting quality objectives into operational processes taking the available resources into account. Quality planning deepens and substantiates corporate planning. Verifiable *quality objectives* for various levels and functional areas of the organisation have to be deduced from quality policies. The quality objectives of an organisation are subordinated to constant verification and adaptation (see Chapter 4.2). Adhering to targets and the possibly necessary cause study and troubleshooting in case of deviations are the tasks of *quality control*. A comparison between the expected specifications and the actual production characteristics (target-actual comparison) provides an important source of information within the framework of quality control. The results obtained form the basis for decisions. Measuring and testing not only relies on the criteria and features of the end product, the measuring and testing data gained from the process are of importance as well (Pfeifer and Schmitt, 2007: 619-659). Quality control is interrelated with *quality assurance*. This is customer-oriented to provide the necessary confidence that quality requirements will be fulfilled.

In the current situation with saturated markets, the guidelines for the individual process steps are characterised significantly by the requirements and expectations of customers (Chapter 2.3). A *customer* may be an (external/internal) organisation or a person (e.g. consumer) that receives a product. The expectations and requirements customers are demanding from a product are constantly changing. Therefore, it is necessary to adapt or increase the level of requirements – mainly with regard to the competitive situation towards other organisations in saturated markets (Chapter 2.2). All measures required for this and the measures to achieve the quality objectives are summarised by the *continuous improvement process (CIP)*. Customer satisfaction is the level to which the requirements have been fulfilled. A positive quality perception creates the basis for differentiating characteristics in the predatory competition (Pfeifer and Schmitt, 2007: 371-385). The quality perceived by the customer is decisive for customer satisfaction. However, consumers quite often cannot differentiate between the terms of quality and grade when they are queried for certain foodstuffs. A *grade* reflects 'the category or rank given to different quality requirements for products, processes or systems having the same functional use' (DIN EN ISO 9000, 2005). It is recommended to state the grade when determining quality requirements. When realising products, a clear distinction has to be made between the product requirements, the product objectives and possibly different quality levels resulting from them. Fulfilling the requirements is nowadays often done by the organisation with different product groups, e.g. the premium private brand, the private brand as standard product and the retail brand. The labels in Box 2.1. printed on the milk cartons reflect different grades but not different quality levels as is often assumed by the consumers.

The requirements demanded from an organisation for realising quality are summarised by the *quality management system* (QM system). The QM system is part of the *management system* composed of many partial areas such as the financial management system and the environmental management system. The *management system* is a 'system to establish policy and objectives and to achieve those objectives' (DIN EN ISO 9000, 2005). For realising quality management systems, industry-neutral and industry-specific standards have been developed in recent years (Chapter 2.2 and Chapter 5). In contrast to the German language which differentiates between norms and standards, in English only one expression, i.e. standard, is used for both terms. Standards correspond to a set of rules which are defined by a special independent expert body. They are audited periodically and, if need be, are then revised. However, the coming into existence, composition and authority of a body between the individual standards is heterogeneous. This excludes the Standards Commission since a specific standard was elaborated for this. Standards are nowadays only rarely designed exclusively for national applications. Meanwhile they refer to a continental application, e.g. Europe, or a global use, to support international trade and to eliminate barriers to trade. The standard designations present the only unique characteristics and thus enable a differentiation. Standards for the agri-food industry are characterised by the abbreviations ISO (International Organisation for Standardisation; international), CEN

Box 2.1. Grades by the example of milk.

- Fresh organic alpine milk with 3.5% fat
- Fresh farm milk with 3.8% fat
- Fresh whole milk with 1.5% fat
- Fresh whole milk with 3.5% fat
- Long-life semi-skinned milk with 1.5% fat

(Comité Européen de Normalisation; European) and/or DIN (Deutsche Industrie-Norm; Germany) and a following number with colon and the year from which the version dates. The family of standards DIN EN ISO 9000 et seq. has been for many years the set of rules providing the framework for establishing and describing a QM system (see Chapter 5).

2.2 Quality management and quality assurance in supply chains to food retail

V. Schütz[1], S. Lehnert[2] and M. Nüssel[1]
[1]DRV, Deutscher Raiffeisenverband e.V., Pariser Platz 3, 10117 Berlin, Germany
[2]University of Bonn, Katzenburgweg 7-9, 53115 Bonn, Germany
schuetz@drv.raiffeisen.de

Quality management is defined according to DIN EN ISO 9000:2005 as 'coordinated activities to direct and control an organisation with regard to quality'. This includes the fields of quality policy, quality objectives, quality planning, quality control, quality assurance and quality improvement. Within an organisation, quality management is the most important link between the different intra-organisational areas such as management, development, marketing, production, sales management or aftersales service and is thus a classical cross-sectional discipline. Quality management is characterised above all by the interaction of customer and supplier. Every process is aimed at fulfilling customer requirements with the emerging product. In particular in recent developments it is necessary to not only fulfil the requirements already known but also the expectations not yet expressed by the customer. Customer expectations have to be determined actively by the organisation. When doing this, the tools and methods of market research, such as surveys, interviews or blind tastings, can be used.

The development process of the agri-food industry can only be fully traced if the term of quality is understood as an important social phenomenon. It is subject to change and social development just like society itself which is using this term within the exchange between business entities. When transferred to the market mechanisms of industry, the socio-economic change will result not only in a shift of the position from supplier to customer, but at the same time the contents which society assigns to the term of quality will also change (Figure 2.1).

In the 1950s, the situation in the various sectors of German industry was characterised by a typical economy of scarcity. Production means, raw materials and technical equipment were scarce so that no demands could be placed on quality preferences. In such distribution markets, the customer is required to consume whatever is offered. Only in markets with increasing market saturation quality will become more important. In the first phase of quality awareness, the level of quality was determined by the supplier. The activities consisted in finding serious defects in the end product and rejecting the relevant products before sale. For better goods, the customer had to pay the increased effort and the higher percentage of waste. According to K.J. Erhardt (personal communication), the position of the customer was characterised in this time by the fact that he was seen by the supplier as an annoying, unavoidable part of the selling process who was to be outsmarted wherever this

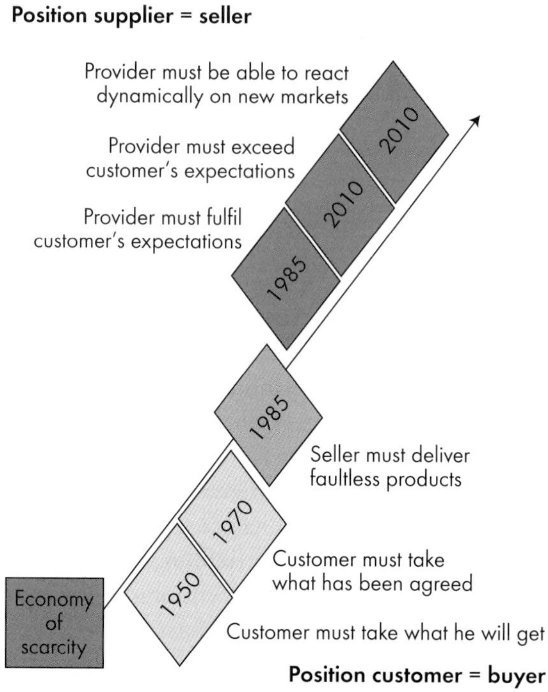

Figure 2.1. Change of the position of customer and supplier (modified from Lehnert, 1998).

was possible. The supplier's management aimed at safeguarding production quantities and if possible increase production by a patriarchal, even authoritarian style of leadership. In addition to the skills acquired during professional training, the employees were not offered any continuous further education opportunities.

In the second phase, production had increased to a level that supply and demand were almost balanced and the market was saturated. With supply increasing, the seller's power decreased, the power of the buyer increased in turn and competition grew. Now, the buyer was able to put first simple demands on quality. He was no longer willing to accept any non-conforming products. In most industries, Acceptable Quality Limit (AQL) values were agreed. These AQL values were not interpreted by the seller as agreed maximum non-conformance rate but as the quality which the customer had committed to buy (K.J. Erhardt, personal communication). The AQL value was misused by the seller whose behaviour at that time was still strongly marked by his already fading market power as permissible non-conformance rate, i.e. the seller made sure that the agreed non-conformance rate was not exceeded or fell below this rate due to sorting measures. Between the parties was an unspoken agreement that the non-conformance rate would not lead to any complaints by the customer within the framework of the agreed limit values. In this phase of 'quality development' it became quickly obvious that systematic measures for quality control were necessary. An enormous sorting effort was met with a small percentage of defects occurred. Only inspection measures within the production process could result in quality improvements. In several industrial sectors, the phase called 'statistical quality control' began (Lehnert, 1994b; Pfeifer, 1993).

Due to the increased competition amongst the suppliers, the customer was able to demand defect-free goods. Non-conformance rates within the agreed AQL values did not lead to returns, however, the customer expressed his dissatisfaction by complaints. At the same time, he expected the supplier to explain how future defects would be avoided. But defect-free goods cannot be guaranteed exclusively by means of statistical inspections methods (see Chapter 4.1). Processes need to be planned, secured and controlled, defects need to be recognised and avoided. Subsequent defect recognition and rectification was therefore replaced by preventive defect elimination (Pfeifer, 1993).

At the beginning of the 1990s, quality assurance systems were increasingly introduced in the companies of German industry. It became rapidly clear during implementation that each and every corporate division could be a potential cause of defects. As from this time on, the entire organisation was included in quality management (Malorny, 1997). In parallel to the quality assurance system, the principle of self-monitoring developed leading to a considerable training effort amongst the employees as they had to assure the required qualification by themselves to be able to assess their activities.

In the third phase, the products became increasingly replaceable owing to their quality characteristics. It was no longer sufficient for suppliers to provide a defined product or service quality to differentiate themselves from their competitors and to retain customers permanently. This had become a matter of course. Now it was important to offer supplementary services to the customer and to fulfil expectations not explicitly expressed. Actual customer satisfaction was less and less dependent on the command of special techniques and processes, but rather contingent on how much each individual employee could identify himself with carrying out his tasks and their importance for the customer. Thus, not only did the requirements change which were imposed from the outside on the companies, but the internal requirements placed on each individual employee changed as well. The performance of all employees in a company together was determined by the personal success of each individual.

Nowadays, it will not be sufficient for the suppliers anymore to fulfil the customer's expectations. Rather they have to exceed these expectations and to surprise the customer time and time again with new, own services before the competitors will do so. Quality management is defining strategies and objectives converting requirements into corresponding products by providing the necessary resources and designing processes to fulfil expectations and customer requirements. In this context, the Aachen quality management model illustrates the intra-organisational interactions and the interactions between customer and supplier within the value chain (see Figure 2.2 and 2.3). A central element of this model is the quality stream focused on recognising and rectifying defects during the individual process steps and in the end product. The continuous improvement process is assured by customer feedback, measurement data of the end product and by data from the development process (quality backward chain) (Figure 2.2).

Implementation in the operative process is done by deducing the requirements and objectives defined at the strategic level. The management level, in turn, is influenced by the customers' requirements, on the one hand, (customer perspective) but also by the motivation to improve the company's success and to maintain the ability of a company to operate (requirements) (Masing, 2007; Schmitt *et al.*, 2007). In this context, three different levels of observation decisive for quality management were defined (Schmitt *et al.*, 2007):

- *Customer perspective:* In the centre of entrepreneurial action is the customer perspective. Taking the economic action and financial success into account by including the available resources, an optimum coverage of requirements and/or product characteristics has

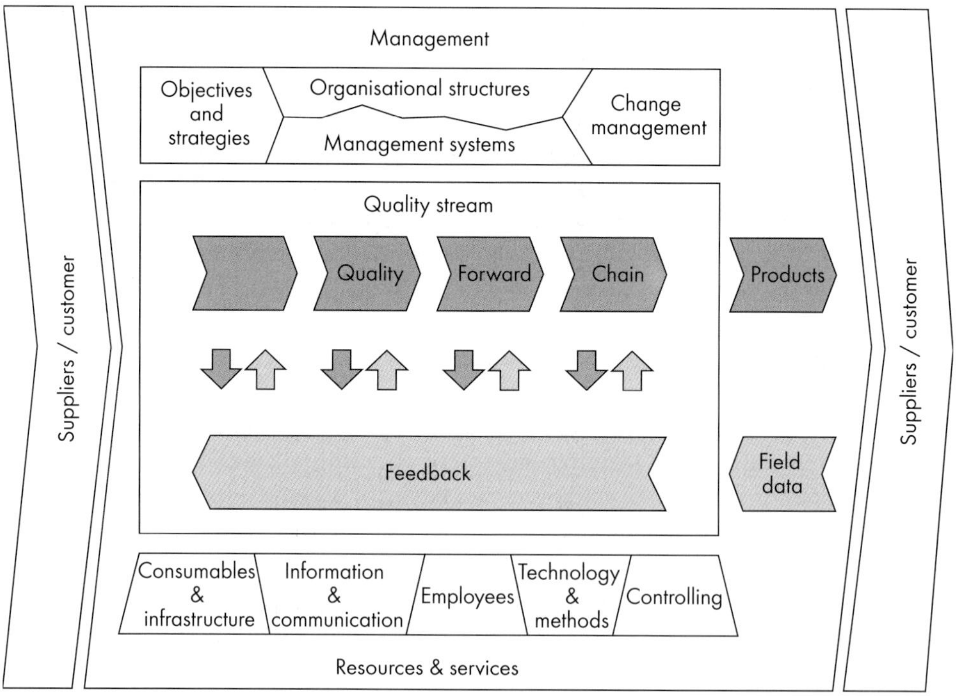

Figure 2.2. The Aachen quality management model (Pfeifer and Schmitt, 2007: 38-41).

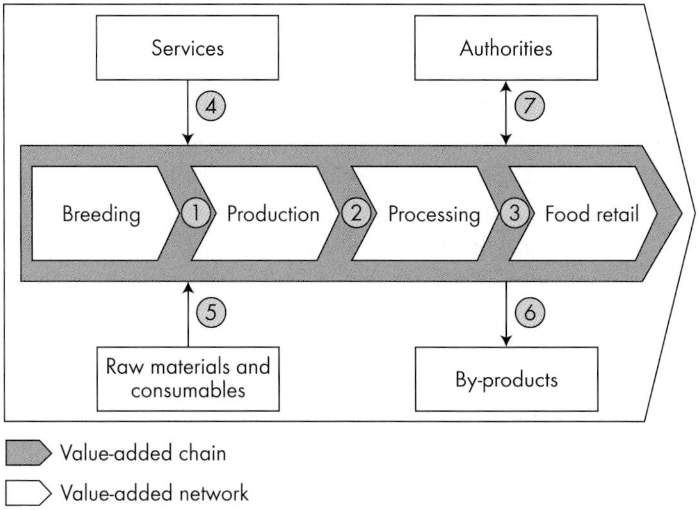

Figure 2.3. Schematic structure of the value chain.

to be achieved. Customer requirements or expectations have to be transformed into the product, verified throughout the production process, measured and improved continuously. The focus of customer perspective is on process and product quality.

- *Management perspective:* The management perspective synchronises the organisation's focus (what is wanted?) with the capacity (what is possible?). It is thus defining objectives for the organisation and is also responsible for the quality policy within the organisation. The focus of the management perspective is top-down and concentrates on system quality and hence on higher-level activities. The management perspective is also responsible for the innovation capability and the transfer of knowledge.
- *Operational perspective:* The operational perspective is responsible for converting requirements from the management and customer perspective and the objectives deduced from them to optimise processes. The limiting factors are the available resources for product realisation. The focus is thus on process quality.

Killich *et al.* (2007) emphasise the importance of employee participation. In the case of the value chain, this could be transferred to the relevant suppliers. The degree of involvement also reflects decisively the success or failure of an intercompany quality management system. The more closely employees or suppliers are involved, the more strongly they can identify with the quality management system. Consequent involvement supports communication between different departments and levels of an organisation sustainably but at the same time cooperative thinking in the value chain.

The production of food is done along the value chain predominantly in collaborative processes with one organisation being able to cover several production steps up to the entire value chain. The degree of integration describes the level of integrated steps of the value chain in one organisation. In food value chains different degrees of integration can be found. The majority of products in the European agri-food industry are produced in collaborative organisations to a large extent organisationally and geographically independent of each other with a lower degree of integration (Lehnert *et al.*, 1995b). Interaction in the production of eggs, poultry and their processing products is characterised most strongly by a high integration level.

The completely different value chains for the production of animal products (eggs, fish, meat and milk) as well as wheat, fruits, sugar, vegetables and wine and the corresponding processing products can be assigned to a common basic structure regarding the sequence of process stages and ancillary processes and can thus be systemised (Figure 2.3).

The decisive stages which can be clearly distinguished from one another are: breeding, production, processing and food retail. The type of variation to be found in the individual areas of responsibility of a value chain illustrated in the figure can be seen in the explanations and examples of Table 2.1. In addition to the four-stage core process, the numbers 4-6 show a secondary area. This encompasses all types of services, raw materials and consumables necessary for the actual value process, by-products occurring during the process, and cooperation with authorities (Table 2.1). Both areas in combination form the value network (see Deimel *et al.*, 2008b; Hoffmann *et al.*, 2008; Lazzarini *et al.*, 2001; Schulze Althoff, 2006; Stock and Lambert, 2001 and Chapter 4.1).

The trade relationships between the core areas 1-3 range from short-term spontaneous trade activities on the spot market up to long-term intensive trade relationships including contractual agreements. Regarding quality management, in addition to the binding national regulations and regulations limited by Union law (see Chapter 3), above all, requirements are characterised by the food retail. These are passed on via processing to primary production. Standards and rules and regulations for quality management put demands in a concrete form and aim at mutual recognition of inspection methods. Certificates support comparability and confidence in the performance of companies. For the agri-food industry, in the past few decades an industry-specific standard (DIN EN ISO 22000, 2005) and various other standards (e.g. IFS, GMP+, BRC, QS) were developed by international bodies – in addition to the universal international standards. The standards are described in detail in Chapter 5.

Food producers are legally bound to implement a HACCP (Hazard Analysis and Critical Control Points) system according to the Codex Alimentarius. HACCP represents an established tool to prevent possible health risks for human beings caused by a foodstuff

Table 2.1. More detailed explanations of the core and secondary areas of the value network.

Area	Explanation	Product examples
1. Breeding	Each foodstuff has its origin in the breeding stage. The objective of this stage is to use the genetic variability to improve, amongst others, animal performance, increase yields, and to generate resistances, by means of targeted cross-breeding and gene technology. The breeding stage provides the genetic source material for the following stages.	Seed, seedlings, sperm, breeding animals
2. Production	Dependent on the foodstuff to be produced, the production stage can be subdivided in subareas, e.g. reproduction, rearing, fattening and slaughtering for meat production. With this stage, agricultural primary production in food production also ends. The products produced at this stage are mostly further processed at the next stage. Only a very small subarea is directly passed on to food retail.	Wheat, fruits and vegetables, sugar, grapes, animals for fattening and slaughter, meat, eggs, milk
3. Processing	During the processing stage, the raw materials are processed and/or packaged as food. These products serve for the consumer as basic products to cook meals or are ready-to-eat convenience products.	Semi-finished and finished products
4. Services	The field of services along the value chain is manifold. These services include consulting, such as feeding and plant production consulting, veterinary stock supervision in agricultural primary production, and consulting for example for the conversion of crisis management systems into processing companies. Another extensive area is the transport of animals and goods between the production stages requiring corresponding background logistics. Many companies use a forwarder to transport their goods. Closely related to this are companies taking over exclusively the trade of goods between the individual stages or combining them with transport. Lab tests and product audits provide another service area. Due to the requirements of quality and health management, various lab tests are necessary to guarantee compliance with primarily legal provisions. Certification of quality management systems has to be mentioned as another group.	Consulting, logistics and transport, lab tests, trade, certification, auditing
5. Raw materials and consumables	Raw materials and consumables are necessary source materials required for the production of the relevant product. These products are needed for the value process, however, they do not result from this process.	Fertilisers, pesticides, feedstuff, animal drugs, spices, packaging, processing materials, flour, dyes, enzymes
6. By-products	These products are substances resulting from the production of the main product. They are often used and processed in other value chains.	Straw, liquid and solid manure, slaughtering by-products, whey, pomace, slop, molasses, CO_2, energy
7. Authorities	Here, control or monitoring of legal provisions, officially recognised audits of industry-specific certifiers and official food monitoring and similar activities are in the foreground (see Chapter 5).	

(see Chapter 5.1). The aim of HACCP is to avoid biological, chemical and physical hazards by operationally adapted critical control points (CCPs) and by control points (CPs) (Andrei, 2005). For this, the process steps within the organisation have to be assessed, and to enable better supervision, management systems have to be established and verified periodically regarding actuality and efficiency (Ellebrecht, 2012; Schillings-Schmitz, 2006; Schmitz, 2006). The principle of prevention is also the foundation for quality management based on voluntary action. The objective of the methods, models, strategies and tools applied within the scope of quality management is to prevent undesired developments or events by preventive activities and measures (see Chapters 4, 8 and 9). When doing this, requirements from legal bases, standards and customer requirements have to be taken into account when establishing and further developing QM systems. The challenges for an organisation or entire value chain consist in structuring and channelling procedures. *Integrated quality management* (IMS) summarises all the existing rules and regulations from different organisational areas (e.g. environmental management, crisis management, safety management) to one set of rules. This approach enables the illustration of the entirety of all processes existing in an organisation, precisely coordinated procedures and responsibilities (Lachenmeir, 2004).

Quality management as an organisational culture is only successful if those responsible also select and apply the correct quality management tools and methods. The latter is only possible if they are known, and are accepted and valued by all actors. Furthermore, the methods provide a schematic framework for the necessary implementation steps.

The excellence model of EFQM (European Foundation for Quality Management) offers help to organisations for establishing and continuously developing comprehensive management systems. The model distinguishes nine criteria consisting of five enablers:

- management;
- strategy;
- employees;
- partnerships and resources; and
- processes, products and services;

and four result criteria *(the results)*:

- customer-related results;
- employee-related results;
- society-related results; and
- key results.

These criteria and their partial criteria are weighted and included as an assessment frame when nominating applicants for quality awards such as the European Quality Award. The company thus has an additional opportunity to improve its image through such awards. The

EFQM excellence model and corresponding prizes create further incentives for companies to orient themselves more intensively towards a comprehensive quality policy.

The following overview matrix (Table 2.2) summarises the most important methods, strategies and models applied in the management of individual enterprises and across several enterprises. When doing this, a differentiation is made between industry-neutral and industry-specific models. The latter are described in detail in Chapter 4 of this book. Application and management fields are assigned to the models presented. In the last column 'Exploration', a reference is made to one or several corresponding subchapters which contain a more detailed description of possible applications.

The basis of a holistic quality management concept is formed by a combination of product, process and employee quality. The social component is decisive here since the personal characteristics of those responsible within an organisation regarding expertise, motivation, ability to work in a team and sense of responsibility are the source for success.

Table 2.2. Overview matrix of methods, strategies and models in different areas of the quality management of individual companies and across several companies.

Designation	Reference			Application					Management areas						Exploration chapter
	Product	Process	Organisation	Quality policy and planning	Product realisation	Quality controlling	CIP	Quality communication	Supplier management	Audit management	Risk management	Crisis management	Energy and environmental management	Corporate social responsibility	
Industry-neutral															
TQM[1,2]	X	X	X	X	X	X	X	X	X	X	X	X	X		2.3, 5.1
Aachen model[1,2]	X	X	X	X	X	X	X	X	X	X	X	X	X	X	2.2
EFQM excellence model[1,2]	X	X	X	X	X	X	X	X							2.2
PDCA cycle[1,2]	X	X			X	X	X								4.2, 5.2
GAP[1,2]		X	X	X		X	X								4.3
FMEA[1,2]	X	X		X	X	X	X	X	X	X	X	X			6.4
HACCP[1,2]	X	X		X	X	X					X	X			2.2, 6.4
Risk portfolios[1]			X								X				6.4
Six sigma[1,2]		X				X	X	X							6.4
Q control cycle[1,2]		X	X			X	X								4.2
Industry-specific															
3-level model[2]		X	X						X	X	X	X			4.1
3A strategy[2]			X					X	X		X	X			4.2
TOF-strategy[2]			X	X		X	X	X			X				4.3
Stage concept[2]			X	X				X	X	X	X	X			4.3
AAD model[2]		X	X	X	X										4.3
Netchain model[2]			X				X	X	X						4.1
Network coordinator model[2]			X					X	X		X	X			4.1
PPP risk management[2]			X					X			X		X		4.3, 6.2, 6.3, 7.1
AMOR[2]	X	X	X			X		X	X		X	X			4.1, 6.4

[1] Applicable to individual companies.

[2] Applicable to several companies.

AAD = Analysis, Assessment, Decision; AMOR = alliances for the mutual organisation of risk-oriented inspection strategies; CIP = continuous improvement process; EFQM = European Foundation for Quality Management; FMEA = failure mode and effects analysis; HACCP = hazard analysis and critical control points; PDCA = plan-do-check-act; PPP = public-private-partnership; TOF = technical-organisational-functional; TQM = Total Quality Management.

2.3 Development of quality strategies in complex customer-supplier relationships

S. Lehnert[1], V. Schütz[2], and M. Nüssel[2]
[1]University of Bonn, Katzenburgweg 7-9, 53115 Bonn, Germany
[2]DRV, Deutscher Raiffeisenverband e.V., Pariser Platz 3, 10117 Berlin, Germany
susanne.lehnert@t-online.de

In the past few decades, not only did the concept of quality change, but with this also the relevant quality improvement concepts. This is a development recognized in all economic sectors, which, however, started in the agri-food industry – referred to the characteristic phases of the principle of quality control to Total Quality Management (TQM) – mostly five to ten years later. The reasons for this are the numerous exemptions, such as the product liability law for the agricultural industry coming into effect only at a later time or legal control activities in companies as part of quality assurance (Figure 2.4).

In recent times, the distance between the agri-food industry and other sectors regarding introduction of improved quality management strategies was shortened. Stronger export focus and further procurement and sales markets capable of development are seen as driving force for rapid adaptation of quality strategies. Figure 2.5 illustrates the direction of development stages in quality control.

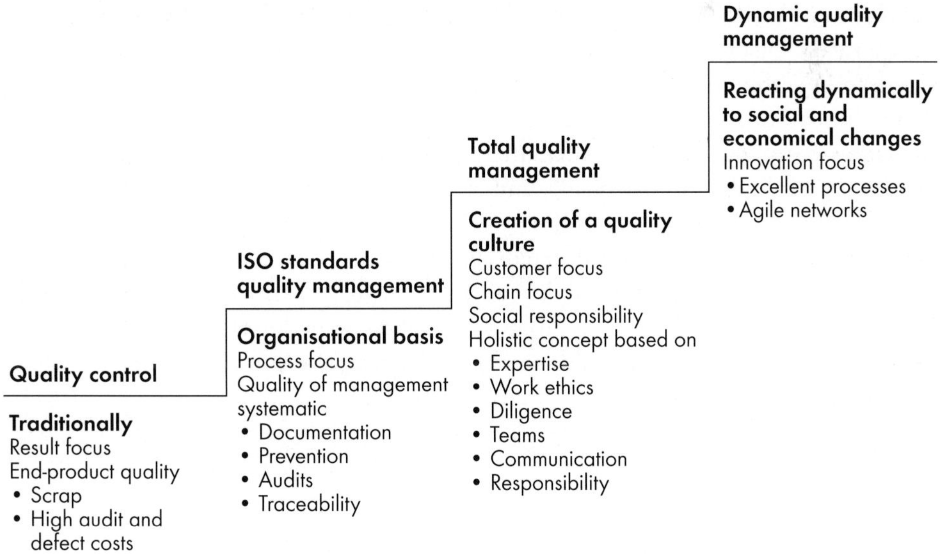

Figure 2.4. Development stages in quality control (modified from Petersen *et al.*, 2002).

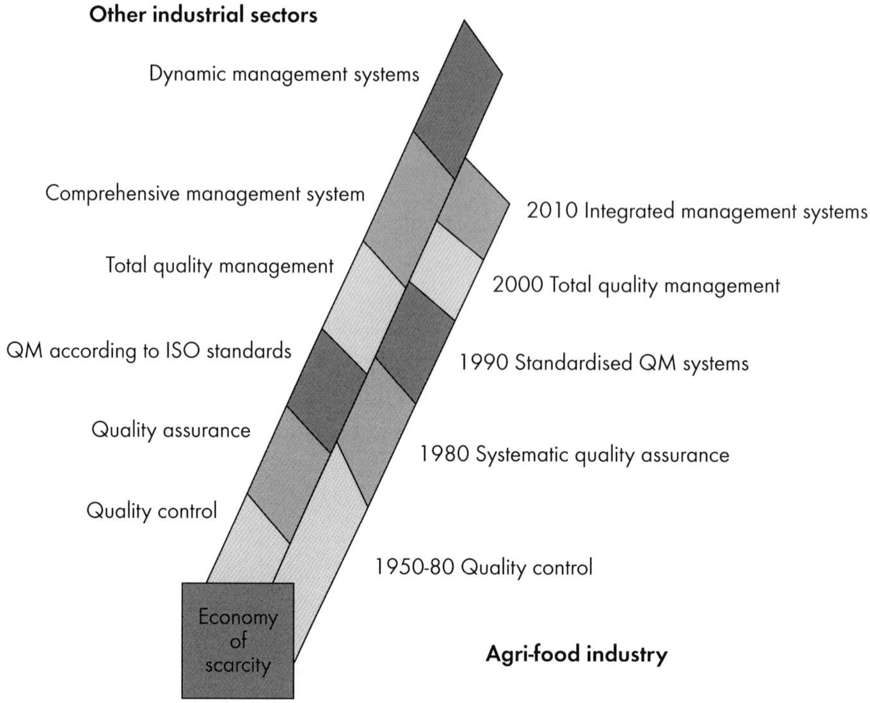

Figure 2.5. Change of the quality assurance concepts in industry and the agricultural sector.

The traditional model of quality control is solely result-oriented and represents a type of sorting process. Quality differentiation only takes place in the end product. The introduction of grades falls within this period. Units not fulfilling the quality requirements are excluded as scrap. In addition, defects are recognised, however, not avoided which leads to high production costs. Frequently, a differentiation is made between three different quality levels: standard goods, medium quality and premium products.

Quality management involving different international standards (see Chapter 5) is concerned with achieving comparability of entrepreneurial performance by systemising. The transparency achieved by this, is promoting confidence in the relevant, often unknown market partner and serves to expand international harmonisation activities and establish business relationships in permanently increasing markets.

The concept of ISO standards is dealing with analysing the management of an organisation and integrating quality more intensively into the business and production processes right from the very start. Thereby, a process-related quality guarantee is to be supported, and the organisation is to outline responsibilities in a way that the previously defined demand on the quality of an end product or service is to be fulfilled and ensured. By recognising the critical parameters on time and by the intervention of preventive and controlling measures during production, errors are to be avoided in advance. Confidence in the efficiency and a good image with market partners is decisive in this regard for implementation and certification of standard-compliant QM systems (see Chapters 4.2 and 5).

The merely technical approach of quality assurance is supplemented by measures to determine customer satisfaction and strategic planning. TQM is designating a still more comprehensive quality concept. With this concept, a culture of cooperation supported by all actors in an organisation is just as much in the foreground as customer satisfaction and social responsibility (see Chapter 8.1). The social component is dominant here, since within a value chain the personal characteristics of the actors regarding expertise, motivation, teamwork, creativity, diligence and sense of responsibility are the source of corporate success and of competitiveness. The voluntary nature of these quality concepts is mandatory with increasing market pressure.

In this context, continuous improvement of quality concepts and agile structures with defined functions, roles and interfaces is necessary to be able to react in the future as well to social and economic changes. This requires controlled and excellent processes to organise and implement quality as the decisive differentiating characteristic in global competition. Nowadays, standards provide a clear framework for organisations in this respect which is audited on a regular basis by certification carried out by neutral third parties. Development and implementation of standard-compliant quality management systems in the German agri-food industry only began in the mid-1990s. In 1987, the International Standardisation Organisation (ISO) with headquarters in Geneva, published the standard series ISO 9000 ff. It presented the first summary and simplification of country-specific rules of more than 100 national institutions in the form of an industry-neutral quality management concept. Two years later, the European Committee for Standardisation (CEN) took over this conception, and only in 1990 the standard was published in Germany. Since then, the family of standards ISO 9000 is regarded as a kind of assistance for 'self-help' in almost all industrial sectors. Food retail and insurance companies also demanded from the companies in the food industry proof of a quality management system. Most of the food-processing companies furnished this proof in the mid-1990s, above all to remain listed by food retailers and discounters important on the market. The advantages for organisations after consequently implementing a QM system based on DIN standards were as follows:

- critical reflection on the own organisation;
- systematic vulnerability analysis;
- optimisation of production and processing procedures;
- rationalisation effects and time and cost savings;
- uniformity, comparability and transparency of results and achievements;
- standardisation of the language used in quality management of all sectors.

Due to the competitive situation, above all large companies in the food industry competing on an international level are forced to present relevant DIN or – similar to their foreign competitors – ISO certificates since most people see them as:

- proof for quality capability visible externally;
- a sign of diligence and quality of company management;
- a signal for extensive control of production processes;
- a confidence-building measure within the customer-supplier relationship.

In contrast to the agricultural industry of the neighbouring countries Denmark and the Netherlands with a strong export orientation, primary production in Germany was very reluctant regarding the use of ISO 9000 standards for developing quality management concepts for the entire value production chain (Anderssen, 1995; Nüssel, 1996). In this respect, mostly regional trademarks, quality labels and certification seals were instead preferred in Germany. At that time, only few pioneers already realized that origin programs would not be sufficient to counter the increased consumer expectations and the rising liability risk. Not only the guaranteed traceability of products in the food chain was demanded, customers also requested guaranteed comprehensible measures of quality assurance within production (Nüssel, 1996).

The Danish conception of quality management in the pork, milk and cheese producing chains was based already in the 1990s – in addition to the framework provisions of the ISO standard – primarily on the formation of cooperating units and a coordinating central quality committee including all the actors directly involved in production.

Essentially, three main objectives were pursued:

- quality leadership in Europe and thus intercompany quality management investing into the future;
- fulfilling market and customer demands for transparency and confidence in product quality;
- safeguarding of export opportunities.

Until today, the quality strategy of a regular flow of information between the individual enterprises, internal and external audits and an exemplary reporting system has managed to achieve an optimum combination of theory and practice and assure in addition to this the advancement of quality standards. The advantages achieved are as follows:

- well organised intercompany conception within the meaning of the TQM approach since all parties affected are involved correspondingly;
- realistic quality policy and efficient production control;
- EU-wide and international recognition and competitiveness.

In the Netherlands, in the 1990s, a strategy in-between a brand program and a quality management system according to ISO standards was chosen with the IKB system (Integrale Keten Beheersing, in English: Integrated Chain Control). The IKB system was designed as a quality management system regulating the form of production along the whole supply chain from the raw material to the consumable end product and guaranteeing the requirements determined beforehand. Participating in this concept is voluntary, but – after the decision to participate was made – certain duties have to be fulfilled. With the IKB system, the Netherlands succeeded in establishing a uniform base level of product quality also aiming at solving domestic problems regarding animal health and the image of meat. The starting points of this IKB concept are customer requests, on the one hand, and rationalisation effects owing to the increasingly targeted cooperation within the value chain, on the other hand (Schouwenburg, 1995).

Against the background of the BSE crisis, in 2001, the QS (Qualität und Sicherheit GmbH) system was established in Germany as a quality assurance concept for the entire chain in which the food retail sector took an active part for the first time. Regarding its approach, the system is comparable with the Dutch IKB system. In this system, the various production facilities of different production stages are bound by contract and obliged to fulfil the specifications laid down by the QS bodies within the framework of quality documentation. Thus, the participants have a rather more passive role. Fulfilling the QS requirements on the part of the companies is verified in addition to the company's own monitoring system in regular intervals by independent certification organisations (Nienhoff, 1995).

The following chapters will describe in detail the development towards establishing quality management systems along entire value chains. The industry-specific quality management activities are always influenced directly or indirectly by the revision processes of the relevant ISO standards. When developing QM systems, companies having only little experience with interpreting the ISO standards so far, often compensate this with more bureaucracy and an increased administrative effort. Standard-compliant quality management usually does not result in much that is new in the production procedure, but it helps to re-arrange well-tried and tested processes and make them more transparent for external parties and for the employees in the company alike. A new and unaccustomed experience is, however, documentation of the operational processes which will be considered in future even more as a necessity due to the product liability situation.

Chapter 3. Legal requirements – with special regard to European and German legislation

Consumers show an increasing interest in the characteristics of their food and demand transparency and credibility for all products offered. These expectations can only be fulfilled by those adhering diligently to legal requirements along the entire production chain.

This chapter will inform you about:
- the essential principles of food law;
- which authorities monitor the legal framework conditions;
- what has to be observed in connection with foodstuff labelling.

Keywords: food law principles, (inter)national legislation, public supervisory authorities, food labelling, designation of origin

Summary

The European Food Law in force lays down central principles for traceability and stage responsibility. Both are drivers for the establishment of intercompany quality management systems in the European agri-food industry. The food law not only contains regulations for measures to protect public health but also for the competition on the international food markets by determining product-specific quality requirements and prevention of fraud. Complexity and multitude of food law issues requires solutions uniform throughout Europe with an interdisciplinary scientific approach.

DOI 10.3920/978-90-8686-789-9_3, © Wageningen Academic Publishers 2014

3.1 General framework conditions: Union law – national law

S. Leible, F. Ortgies and S. Schäfer
University of Bayreuth, Department ZR IV, Universitätsstraße 30, 95447 Bayreuth, Germany
stefan.leible@uni-bayreuth.de

The objectives of the food law are, on the one hand, protection of human health, and, on the other hand, protection of the consumers against fraud and deception. The legal requirements necessary for this are complex. In addition to national regulations, in particular, the regulations of the Union (i.e. European regulations) have to be considered.

A distinction has to be made between national law and Union law. At national level, furthermore, a differentiation has to be made between federal and state legislator. Both regulation levels consist moreover of formal and material laws. *Formal laws* are enacted by the bodies of the legislative mentioned in the state or federal constitution, *material laws* (regulations) are, however, enacted by the government or public authorities (executive). For this, an enabling provision within a formal law is necessary.

Union law is divided into primary and secondary law. *Primary law* designates, in particular, the law of the European Treaties (Treaty on the Functioning of the European Union, TFEU; Treaty on European Union, TEU). These treaties rank highest within the standard hierarchy of Union law. *Secondary law,* often also designated as derived law, is the law enacted by the bodies of the Union. Above all, regulations and guidelines are the typical form of action. The Union *regulation* (Article 288 Para. 1 TFEU) has immediate effect in each member state. Transposition into national law is neither necessary nor admissible. The member states have to make sure that the regulation is applied without any limitation. In individual cases, additional implementing provisions of the member states or implementing regulations of the European Union might be necessary.

The second type of action is the *directive* (Article 288 Para 3. TFEU). It has to be transposed into national law by the member states so it can apply. Regarding transposition, each member state can freely decide on the means and form of transposition within the limits allowed by the directive. The guideline on its own does not establish any rights for individuals. In exceptional cases, it can apply directly, amongst others, if the national legislator has allowed the transposition deadline to expire without acting upon it or if transposition was faulty.

Union law has *priority* over national law. This priority in application of primary and secondary law is effective vis-a-vis all national regulations. In addition to this, these have to be applied so that they are compatible with Union law and give effect to this law in its entirety.

An important component of the European food law is the regulation laying down the general principles and requirements of food law, establishing the European Food Safety Authority and laying down procedures in matters of food safety (EU, 2002), in short General Food Law Regulation. It establishes a general part of the European food law and ensures a protection level equally high throughout Europe. The main concern of the General Food Law Regulation is food safety. It provides the foundation of various further regulations and directives regulating food-law issues.

The cornerstone of the food law in Germany in general is the Food, Articles of Daily Use and Feed Code (LFGB). The regulatory areas of the General Food Law Regulation and the LFGB are closely interrelated. The LFGB is closing some of the gaps left by the General Food Law Regulation, on the one hand, and on the other hand it refers to already existing rules in the General Food Law Regulation to avoid a repetition of standards.

3.2 Principles of food law

S. Leible, F. Ortgies and S. Schäfer
University of Bayreuth, Department ZR IV, Universitätsstraße 30, 95447 Bayreuth, Germany
stefan.leible@uni-bayreuth.de

The protection of human beings against foodstuff hazardous or harmful to health is the original task of the food law and also characterises its basic principles explained in further detail below.

Traceability

The principle of traceability is especially a consequence of the BSE crisis. Accordingly, full *traceability* (Article 18 of the General Food Law Regulation) along the entire value chain is mandatory. ('From farm to fork', Article 3 No. 15 General Food Law Regulation). Each food business operator must document his buyers and suppliers so that forward and backward traceability is guaranteed at all production, processing and selling stages ('one step up, one step down'). This traceability is necessary for damage control so that in case of damage, targeted and rapid access is possible. For this, it is indispensable, however, that all participants within the production chain without any exception comply with their documentation duty. In addition to this, unhindered data flow is necessary to make sure that the relevant information reaches the right addressee (e.g. supervisory authority) quickly in case of damage.

The documentation duty ordered by the General Food Law Regulation does not only apply to in-house documentation. Such documentation is, however, recommended for each company and is mostly guaranteed by corresponding QM-systems. The General Food Law Regulation thus only assures a minimum standard of traceability which can be extended (on a voluntary basis). Within the production chain, a company-specific quality assurance system is provided normally as the basis for contractual relationships which means that often at least a factual access exists to establish such a system.

Stage responsibility

The documentation obligation correlates with the so-called *stage responsibility* of the food entrepreneur (Article 17 and 18 of the General Food Law Regulation). Each food entrepreneur is responsible in his processing or distribution stage to ensure safety in terms of food law. The entrepreneur is liable for the safety of the product and must assure that no harmful changes occur. In addition to this, comprehensive responsibility of the various service providers involved along the production and distribution chain also has to be mentioned. In particular, the increasingly growing sector of food delivery via internet raises numerous issues in this respect. Therefore, it is discussed at the moment if such a delivery agent is a food entrepreneur and thus subject to the far-reaching system of obligations. This has to be answered in the affirmative based on the above mentioned assumption.

The principle of mutual recognition

The beginnings of the food and feed law of the EU were characterised by the objective of *full harmonisation*. Establishing a uniform law failed, however, due to the multitude of areas to be controlled. The form of action of the directive which was initially the main instrument of the European food law turned out to be unsuitable. Therefore a paradigm shift was made. The objective of full harmonisation was relinquished owing to the landmark decision of the European Court of Justice 'Cassis de Dijon', especially since it became clear that standardisation throughout the Union would not do justice to the cultural variety in Europe. Since then, the 'Cassis de Dijon' principle has stood for the determinant maxim of mutual recognition. The rules of the member states regarding production and marketing of food are considered as equivalent in principle. Therefore, a product legally marketed in one member state is basically marketable in the whole of the European Union.

Principle of misuse and prohibition

The food law is characterised significantly by the so-called *principle of misuse*. A manufacturer is allowed to market any foodstuffs produced by him if he is not overstepping the mark provided by laws and regulations. Preventive approval or authorisation is not necessary. An important limit is for example the prohibition to market unsafe and thus unsuitable and dangerous foodstuffs.

For some areas of food production and food sale, classified by the legislator as especially sensitive and requiring monitoring, the so-called *principle of prohibition*, however, applies (with conditional approval): according to this principle, it is prohibited to market a foodstuff if this has not been expressly allowed beforehand. The principle of prohibition applies, for example, to food additives, flavours and enzymes. These can be used in foodstuff only if they are contained in the positive lists of the corresponding regulations. The so-called health claims regulation (EU, 2006) has extended the scope of such a principle of prohibition, for the first time, from the composition of food to include labelling and advertising of food.

3.3 Monitoring, risk assessment and prevention

S. Leible, F. Ortgies and S. Schäfer
University of Bayreuth, Department ZR IV, Universitätsstraße 30, 95447 Bayreuth, Germany
stefan.leible@uni-bayreuth.de

Adhering to legal framework conditions is monitored by public authorities. They rely increasingly on self-commitment and self-control of the food industry. Public surveillance in terms of food law is limited quite often to 'controlling the control'. Only in those instances where this is not possible or insufficient, the government is making an appearance as an actor carrying out more comprehensive controls.

Structure of administration

Germany is structured according to the federal principle. The constitution provides that normally the authorities responsible in the states implement food regulations (Figure 3.1). These are, primarily, communal authorities, i.e. trade supervisory boards or veterinary inspection offices. The repertoire of the official competence to act ranges from complaints and official warnings up to closing down of companies (see Food and Feed Code: §§ 39 ff.). In addition to this, there also exist, naturally, important authorities at federal and European level.

Federal Office of Consumer Protection and Food Safety

The Federal Office of Consumer Protection and Food Safety (BVL) located in Berlin coordinates the monitoring activities of the states in the area of food safety. For this, it determines those fields where food and feed crises of nationwide dimensions can occur. Subsequently, strategies for avoiding and handling risks are developed. This involves first of all, risk management measures which ideally should be taken by all authorities responsible

---

Box 3.1. Risk factor of online food trading.

At the moment, the Federal Office of Consumer Protection and Food Safety is especially concerned with online food trading. In connection with this, it has identified certain significant risks, i.e. trading with potentially dangerous food additives from non-European countries in Germany. In order to minimise the risks determined, the BVL has to rely on the cooperation of the regional authorities since in most cases only they have the corresponding powers, e.g. to declare a ban on engaging in commercial activities, etc.

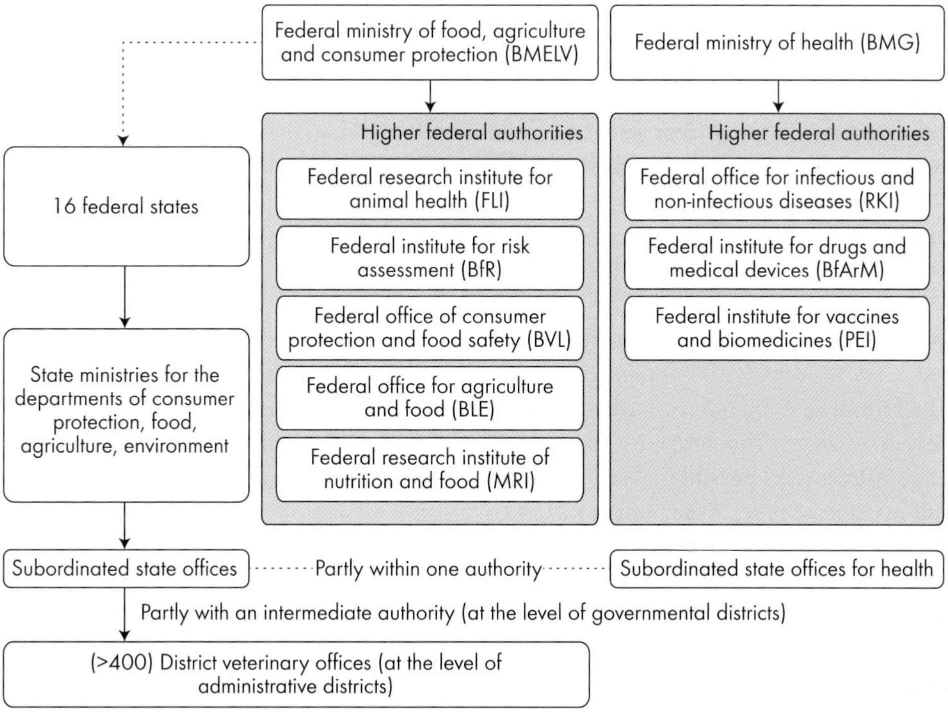

Figure 3.1. Responsible federal and state-level ministries and administrative structure in health-related consumer protection (figure by Tilman Wilke).

in the states to avoid certain risks materialising. If despite this, a food crisis develops, the BVL will apply the management strategies developed beforehand.

This Federal Office also has, in addition to this, an important interface function in relation to the European Union. The EU is maintaining a rapid-warning system, the so-called 'Rapid Alert System for Food and Feed' (RASFF). If German authorities realise that a foodstuff traded here is dangerous, the BVL is responsible for entering a corresponding message in the RASFF so that other member states are also made aware of the danger and can take measures for the protection of health.

Federal Institute for Risk Assessment

Already the name of this Federal Institute for Risk Assessment (BfR), also based in Berlin, leaves no doubt about its statutory mandate: the BfR assesses scientifically the (health) risks posed by a foodstuff or substance. For this, it undertakes corresponding researches and prepares reports and recommendations. The political and administrative decision-makers can take legislative or administrative measures on the basis of these findings. This makes apparent as well the strict separation between (scientific) risk assessment and (political executive) risk management (Figure 3.2). The BfR has the greatest possible institutional independence to avoid that scientific determination and analysis of risks is in danger of becoming politically influenced.

In addition to scientific risk assessment, the BfR is also obliged to communicate its research findings to the broad public (risk communication).

European Food Safety Authority

The European Food Safety Authority (EFSA) plays a similar role at European level as the BfR in Germany. This authority, created as part of the reorganisation of the European Food Law, assesses risks possibly caused by foodstuff but has no political enforcement power in risk management according to the law.

Since its foundation, many new tasks have been transferred to the EFSA. This authority is for example responsible for scientific assessment of health-related information according to the Health Claims Regulation. Due to the vast number of information to be verified, this task brought and brings this authority to the limits of its efficiency.

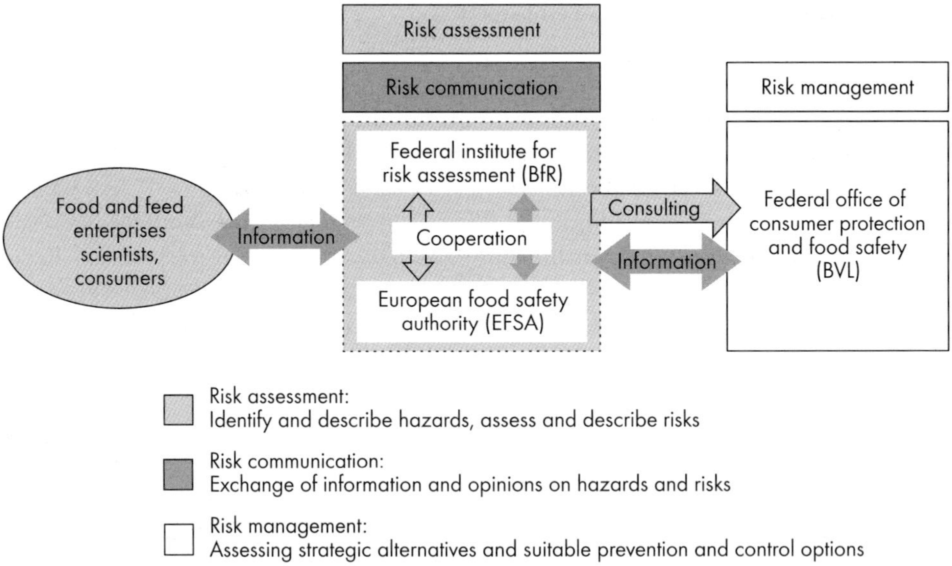

Figure 3.2. Process of risk analysis at national level (modified from Schmitz, 2006).

3.4 Labelling of products and designation of origin

S. Leible, F. Ortgies and S. Schäfer
University of Bayreuth, Department ZR IV, Universitätsstraße 30, 95447 Bayreuth, Germany
stefan.leible@uni-bayreuth.de

The object of protection of the food law is the consumer. The intention is not just to protect him against health risks, but he is to be enabled to take reasonable purchase decisions on the market. This is an indispensable prerequisite for effective competition – this also applies to the food sector.

The consumer must be precisely informed about the characteristics of a foodstuff important for him to be able to speak about real freedom of choice. This is ensured by the legislator in two ways, i.e. he obliges companies, on the one hand, to provide certain information. The food law contains a multitude of obligatory labelling elements. On the other hand, the legislator prohibits food enterprises as well to deceive customers by misleading information about the characteristics of the products offered.

Depending on the level of protection required by the relevant customer group, the information requirements increase or decrease: commercial re-users of a foodstuff are less in need of protection than the end user. Differences also result from the fact if a foodstuff is sold ready-packed or as a loose good.

Obligatory labelling elements for ready-packed products

The majority of the obligatory labelling elements to be provided on the packaging by the manufacturer are determined at the moment still according to German law but with the coming into force of the European regulation on the provision of food information to consumers (EU, 2011) as from December 2014 solely according to European law.

In particular, the so-called sales name (according to the regulation on the provision of food information to consumers in future only 'name') is important. This name refers to the name of the foodstuff by which the customer can identify the product.

The inclusion and design of a list of ingredients also has been prescribed in detail: the ingredients to be mentioned but also the order in which they are given is not up to the manufacturer but legally prescribed. In general, the ingredient representing the largest proportion of the foodstuff has to be mentioned first. The other ingredients follow by their order of weight. Under certain conditions, their percentage in the food has to be stated in the list (QUID declaration) in connection with individual components. Allergens have to be identified especially.

It is always necessary to indicate the date of minimum durability and in case of perishable foods also the use-by date.

Nutrition labelling up to now still voluntary, is obligatory in future according to the regulation on the provision of food information to consumers ('Big Seven'). Already now, many manufacturers provide a corresponding table on their products. The proposal to introduce a 'traffic light labelling' to show the contents of fats, carbohydrates and proteins could not be enforced. Instead of this, a system with a daily consumption recommendation is preferred.

The type and form of labelling, namely font size and position of the relevant elements, is not up to the relevant company alone. Legibility of the information must be assured, which will be quite a challenge for every packaging designer from 2014 owing to the multitude of obligatory elements mentioned here only by way of example and the prescribed minimum font of 0.9 mm especially for small packages, but not only for these.

'Organic' and 'without gene technology' – labelling according to the Council Regulation on organic agriculture

Organic products are trendy. Therefore it is quite attractive for the food industry to place the attribute 'organic' on their products. In order to avoid misuse and prevent differing term concepts, the EU has already enacted in 1991 a first regulation on organic production of agricultural products and labelling of organic products replaced in 2009 by the new EC basic regulation on organic production (EU, 2007).

This basic regulation on organic production contains comprehensive production regulations for products from organic agricultural production. The use of fertilisers is for example prohibited, the production methods are also described in detail. A closely monitored control system guarantees that the rules are complied with.

Products may bear labels referring to organic production only if the requirements in the product-relevant part of the regulation are adhered to. Article 23 of the basic regulation on organic production is controlling the exact form of this. If a foodstuff corresponds with its requirements, it may use terms such as 'organic' or 'ecological'. It is, however, not admissible to use a term for a certain foodstuff which induces a production conforming to the regulation although this is not actually the case.

The EC basic regulation on organic production is furthermore determining the so-called union logo, a stylised oak leaf consisting of white stars on a green background. This is to be attached on ready-packed foods if they comply with the provisions of the regulation. It is also possible to use other seals assigned by private certification bodies. Quite often, stricter requirements apply to their use. Sometimes, they also have a different meaning than the seal based on the basic regulation on organic production. Their use, may, however, under no circumstances mislead the consumer (Chapter 3.5).

Labelling 'without gene technology'

The German consumers' request for products possibly free of gene technology conflicts with the fact that meanwhile it has become rather difficult to produce certain raw materials, e.g. soya, in a quality absolutely free of gene technology. This circumstance has persuaded the German government to regulate the labelling 'without gene technology'. In doing so, it was confronted with the challenge to establish a trustworthy system for consumers but keeping at the same time the access obstacles for the food industry as low as possible. It has not been successful in either of these plans.

It is, above all, quite problematic in connection with the negative labelling in force (= 'without gene technology') that it allows certain contacts of a foodstuff with genetically modified organisms (GMO). For example, animals need not be fed throughout their live with GMO-free feedstuff; rather, adhering to certain waiting periods it is sufficient to still use the label 'without gene technology' for a product. Studies show, however, that the consumer indeed expects in conjunction with the statement 'without gene technology' not only that the production is mostly free of genetically modified organisms but completely free of them. Since the legal position in Germany does not correspond with this conception it must be urgently reformed.

Geographical indications

The protection of geographical indications is guaranteed in Germany in particular by §§ 126 ff. Trademark Act. If there are any gaps in the protection, it is additionally possible to refer to § 5 UWG (Unfair Competition Act) and § 11 Para. 1 p. 2 Food and Feed Code. The protection offered by the trademark act is, however, more specific so that this prevails basically.

A distinction has to be made between direct and indirect geographical indications. The direct indication characterises a foodstuff not directly by the use of the geographical term; the consumer, can, however, conclude the geographical origin of the foodstuff, e.g. owing to the use of (country) specific symbols or colours. In contrast with this, the indirect geographical indication represents a geographical term related with the good ('Aachener Printen', i.e. gingerbread from Aachen, 'Lübecker Marzipan', i.e. marzipan from Lübeck) and which can be directly associated by the consumer with the special characteristics of the product. The geographical origin thus gives the foodstuff its typical characteristics.

However, a certain protection is also given at Union law level. Regulation 1151/2012 (EU, 2012a) offers protection for geographical indications and designations of origin through registration in a list applicable for the EU. In this respect, a distinction is made between indications of origin – the foodstuff gains its characteristics from its specific geographical nature and is exclusively produced here – and mere indications of provenance.

The protection is guaranteed to avoid fraud and misleading of consumers and to prevent that the partly considerable economic value of the indication of origin is endangered by dilution. The key difference to the trademark is that the geographical indication of origin regularly is not an individual right but can be used by several authorised parties.

3.5 Prevention of fraud

S. Leible, F. Ortgies and S. Schäfer
University of Bayreuth, Department ZR IV, Universitätsstraße 30, 95447 Bayreuth, Germany
stefan.leible@uni-bayreuth.de

Conception and implementation

The objective of the law on fair trading practices is, amongst others, the protection of the consumer against fraud. Therefore, above all the Unfair Competition Act provides rules for all companies for the 'battle for customers'. Complementary to this, the food law contains in § 11 Food and Feed Code and in Article 16 of the Basic Regulation autonomous facts prohibiting misleading consumers. According to this, it is, above all, prohibited 'to commercially market food under misleading designations, indications or layouts'.

Misleading advertisement and labelling are very harmful – for consumers as well as for competitors. In most cases, it is, however, not possible to determine in general but only in each individual case what misleading actually means.

Here, the Unfair Competition Act provides the possibility to separate food standards from the area of public law addressing them within the framework of proceedings under private law. This implicates that no longer only public authorities can prosecute fraud, but also competitors and special consumer associations. This will improve protection for all market participants and relieves the state from its function as market supervisor.

Assessment scale for fraud: customary usage in the trade

What is fraudulent and misleading is judged by courts according to the circumstances of each individual case. The so-called *customary usage in the trade* provides the basis for the assessment criteria to determine if a label is misleading the consumer. To determine this customary usage in the trade, the relevant court tries to find out what the consumers deduce from a certain indication and if this perception corresponds with reality. If judges find that perception and reality diverge, they mostly affirm the fraud.

However, it still needs to be clarified which ideal consumer is taken as the basis for assessment, i.e. whose position the judges support when assessing a statement. In this respect, the European Court of Justice (ECJ) has shaped the model of a reasonably well informed, attentive and sensible average consumer. This consumer is, above all, characterised by the fact that he can understand the information provided on a packaging, e.g. the list of ingredients, and is intellectually able to interpret this information correctly.

To determine the customary usage in the trade, the court can make use of various aids. Mostly, it is satisfied with determining the customary usage in the trade by itself. This is always possible when the members of the relevant bench of judges belong to the service categories addressed by the labels. This is almost always the case with food. Furthermore, it is possible to actually carry out consumer surveys and seek relevant expert opinions. In practice, this seldom occurs, however, due to the cost and unpredictability involved.

Guidelines of the German Food Register

In this respect, the guidelines of the German Food Register have a special importance. This register contains descriptions of the customary usage in the trade relevant at a certain point of time sorted according to different classes (such as for example honey and honey products or meat and meat products). The guidelines work like anticipated expert opinions which can be referenced during proceedings. The court is not bound by them, it will, however, only rarely deviate from their contents since the guidelines are constantly audited and revised by an expert commission with more than 30 members consisting of various representatives from the food industry, consumers and academia.

Fraud regarding quality or quantity

According to § 11 of the Food and Feed Code, in connection with labelling of food as well as advertising food, misleading references to the nature in particular are prohibited. This includes for example indications on the quality of a product. Quality-enhancing indications such as 'selection', 'singular', 'precious', etc. may only be furnished if they are true, i.e. if the quality of the food is actually above average.

Indications regarding freshness can also be misleading in connection with the quality of foodstuffs. It would be for example misleading to market pasteurised whole milk as 'fresh whole milk'. Also misleading are false quantitative indications on a packaging, and this fraud can be determined quite simply.

Fraud regarding operational origin

An indication of operational origin is provided if they impart to the consumer groups addressed an idea of production and preparation of foodstuffs taking place in a certain enterprise or certain type of enterprise. Examples for this are the indications 'monastery beer' or 'cellar beer'.

Advertising with naturalness

If characteristics of a product are emphasised which are typical for other comparable products, this can be misleading for the consumer. He could wrongly assume that only this very product is in possession of the characteristics mentioned without recognising that all other foodstuffs as well have the same features. This wrong perception could influence his purchase decision.

Imitated foodstuffs

The group of the so-called 'imitation foodstuffs' has time and time again made the headlines ('analogous cheese'; 'imitated ham'). The food law has always prohibited marketing imitated food if this can lead to confusion with the original. Careful verification is always necessary in this respect; because a confusion risk on its own can basically be prevented by adequate labelling. The scope for determining such a confusion risk in turn provides the model of a considerate average consumer also taking the list of ingredients into account when assessing a product.

Indications of purity or nature-related indications

Indications such as 'natural', 'naturally pure', 'pure', etc. do not signal to the consumer that the products have been produced organically within the meaning of the Organic Basic Regulation. These terms are rather associated with the nature of the product itself. The consumer expects that products labelled correspondingly are free of residues and additives. According to the so-called 'D'Arbo Decision of the European Court of Justice' (Reference C-465/98) one cannot, however, assume that a suitably informed and reasonable average consumer is expecting even with terms such as 'natural', etc. that the relevant products do not contain any residues. He will rather bring to account that with a great number of foodstuffs (ubiquitary) smallest traces of residues present everywhere can be contained in the food. Hence, he does not expect that the food is completely free of residues but mostly free, i.e. assumes naturalness.

Chapter 4. Quality sciences as cross-sectional discipline

Quality management as a scientific cross-sectional discipline in the agri-food industry requires in addition to a stable theory also findings from neighbouring management disciplines. This chapter will present the most important models and strategies in intercompany quality management.

This chapter will inform you about:
- which models illustrate general interrelationships and the roles of actors in quality and risk management,
- which procedural models can be used for planning of quality, environmental and crisis management systems.

> *Quality science is about collecting experience, systemising findings and searching for regularities and methodical procedures as well as defining uniform terms in the field of quality management (GQW)*

Keywords: intercompany quality management, procedural models, risk management, quality control cycles

Summary

Substantiating quality management theoretically is necessary to understand general processes and industry-specific organisational structures better and classify scientific models more easily. In the agri-food industry there exist horizontal networks of companies at the various stages of the value chain offering the same products and being interrelated. The Netchain model is best suited to abstract the various vertical and horizontal transaction relationships within the production cycle. Consolidating and processing of quality information along the Netchain, requires the willingness to cooperate and the existence of coordination services. In this respect, network coordinators have a particular role. They do not only control and regulate intercompany communication and service processes in quality management but also coordinate the exchange of information in crisis management

DOI 10.3920/978-90-8686-789-9_4, © Wageningen Academic Publishers 2014

between public organisations and companies along the value chain. The design of intercompany information and communication systems is an essential precondition to initiate a continuous improvement process as an overall objective of quality management.

4.1 Models in quality management

B. Petersen and J. O'Hagan
[1]University of Bonn, Katzenburgweg 7-9, 53115 Bonn, Germany
b-petersen@uni-bonn.de

The following sections introduce essential models in quality management of the agri-food industry. Nine graphics illustrate the models and interrelationships between customers and suppliers, and between private and public organisations.

The process model

An organisation establishing a standard-compliant quality management system and seeking to be certified for example according to DIN EN ISO 9001 (see Chapter 5.1), must show how it is organisationally guaranteeing continuous improvement. From an organisational point of view, the demands for continuous improvement seem to represent a contradiction regarding the structure-conservative image of an enterprise as a system endeavouring to remain stable. Continuous improvement requires, however, certain dynamism. A continuous improvement process (CIP) hence needs constant involvement of all actors of an organisation and communication of the actors amongst each other, since otherwise objectives are not realised and the entire improvement process will stagnate. Those responsible for QM therefore have to invest resources (money, personnel, and time and energy) in a targeted form to introduce and maintain a CIP. Behind these necessary improvement processes, there are three experiences made time and time again:

- CIP supports flexibility of an organisation;
- CIP facilitates adaptation to changing markets;
- CIP is the motivation from the inside – i.e. from an attitude of everybody involved in a value chain – to align an organisation towards changes.

If everybody involved assimilates this internal attitude to continuously improve their own achievements, some or all of the following positive effects materialise:

- the organisation saves on resources and discovers synergies;
- it improves procedures and optimises processes;
- it improves material products or services and increases customer satisfaction;
- it reduces waste and cuts cost;
- it initiates, optimises and implements innovation processes;
- it supports creativity and the commitment of employees and members;
- it improves the company culture and teamwork;
- it reduces the time and performance pressure on employees, service providers and suppliers.

The process model represented schematically in Figure 4.1 illustrates the great number of different measures which can be used by quality management to initiate and maintain a continuous improvement process:

- take over responsibility;
- measure, analyse and improve processes;
- realise products with required characteristics and features;
- manage resources targeted to the quality policy;

Individual enterprises but also groups of companies must recognise and control several interconnected and interrelated processes to be able to successfully work with a QM system along a value chain. In this respect, the following categories of processes can be distinguished:

- *Core processes* are the most important business processes for the success of a company or group of companies. They can be deduced from the relevant core competencies of a stage within the value chain (Gerboth, 2002; Pfeifer, 2001; Thomas, 1990) (for example rearing of new-born piglets up to the selling age and sale to fatteners) illustrated in Figure 2.3.
- *Support processes* have no directly recognisable use for the external customer and are not directly involved in the value chain. They support the core processes and run mostly in parallel to these (Gerboth, 2002; Griese and Sieber, 1999; Pfeifer, 2001). They also include the manifold services in quality management used in particular by agricultural enterprises (e.g. crisis management, data management, veterinary stock supervision).
- *Management processes* include strategy, organisational and budget planning activities of the top management of an organisation. They mainly refer to longer periods than support processes (e.g. marketing).

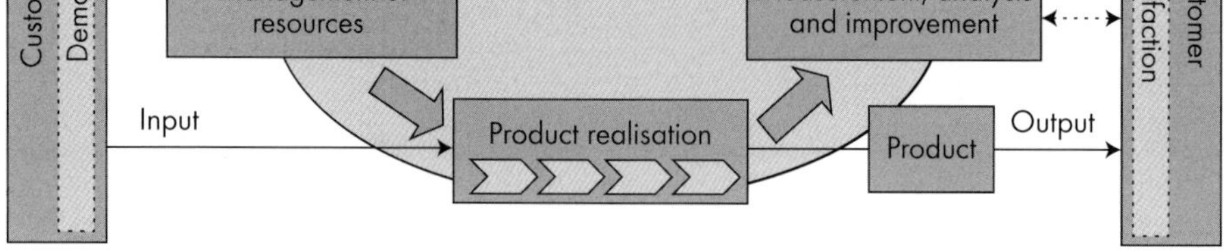

Figure 4.1. Process model of DIN EN ISO 9001:2008.

Recognising and systematising these different processes and taking the process interrelationships into account is called 'process-oriented approach'. Quality in the intercompany interaction of actors is understood here as well as the optimum fulfilment of customer requirements. In this case, it is a network of internal and external customer-supplier relationships (Pfeifer, 2001). In the introductory Chapter 2, the Aachen model and the Munich model have already been mentioned, both following the approach of the process model of DIN EN ISO 9001:2008. This also applies in principle for the process model for value chains in the agri-food industry defined by Schulze Althoff (2006). His industry-specific process model focuses as well on the manner in which advance information and feedback between customers and suppliers is arranged. The goods exchange and thus the business processes are accompanied by a systematic exchange of product-supporting information. In Chapters 8 and 9, in particular the experiences made when organising and coordinating intercompany systems is addressed by way of examples.

Figure 4.2 describes his model as a sequence of inspections to determine information to uniformly control the core processes of the individual stages of a value chain. Here, the following is distinguished:

- *incoming inspections* referred to the individual supplier;
- *intermediate inspections* referred to the relevant producing company itself; and
- *outgoing inspections* referred to the relevant customer.

If all specialised enterprises in a value chain follow this sequence of inspections referred to their core process, the information relevant for decisions deduced from them assures the

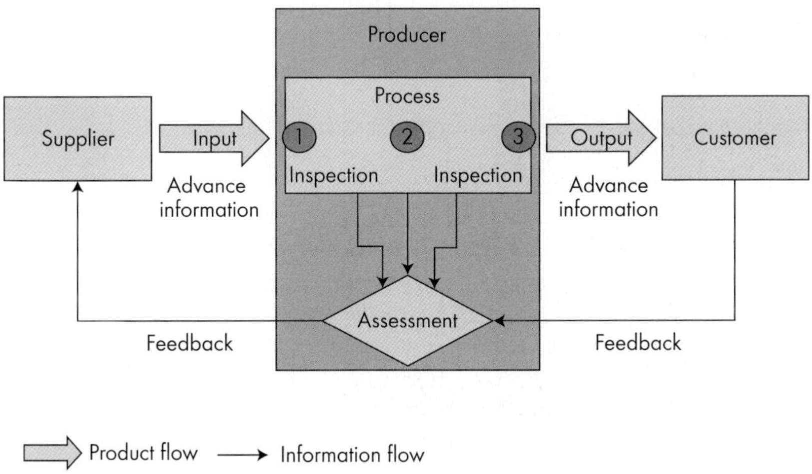

Figure 4.2. Process model of goods and information exchange between suppliers and customers (Schulze Althoff, 2006).

process and product quality of the goods sold demanded by the last distribution stage (e.g. food retail or catering trade). Schulze Althoff (2006) combines in his approach the process model derived from DIN EN ISO 9000 with the Netchain model defined by Lazzarini *et al.* (2001).

Figure 4.3 illustrates the exchange of goods and information within the chain and the specific position of each individual enterprise in its role as a supplier but also as a customer within the value processes. This special aspect of the exchange of inspection information is illustrated in more detail later on when explaining the AMOR model.

The netchain model

Regarding the processes of the models illustrated here to describe the value chains, two aspects have to be taken into account. On the one hand, these are processes determined biologically (such as reproduction and growth), and on the other hand, technically (such as rearing, production, processing and transport). The organisational prerequisites for customer-supplier relationships and the layout of business processes present major differences while, meanwhile, the biologically and technically determined processes are similar throughout

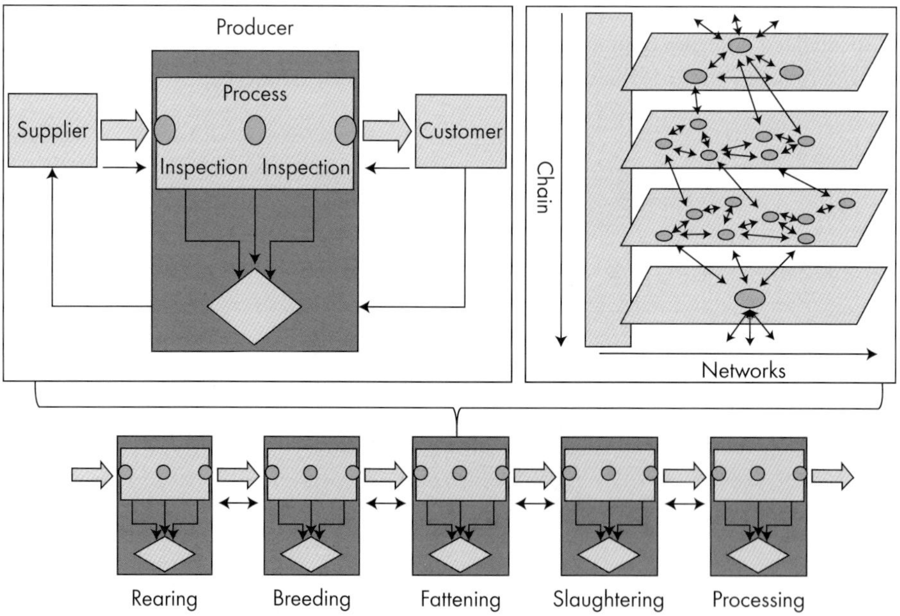

Figure 4.3. Model for the exchange of goods and information in value chains (Schulze Althoff, 2006).

the world. In this respect, manifold forms of cooperation and integration in networks and chains (Horvath, 2004; Schulze Althoff, 2006; Spiller *et al.*, 2005; Trienekens and Beulens, 2001b) (Chapter 8.2) are described and analysed.

An example for a generic model is the 'Netchain' which tries to combine the characteristics of networks with those of value chains. According to Lazzarini *et al.* (2001), a 'Netchain' (a neologism created out of the terms of network (net) and (chain)) consists of the links of horizontal enterprise networks interconnected vertically (along the food chain). This Netchain encompasses even two sectors in the case of the agri-food industry (Figure 4.4). This type of link between the two sectors is typical for the value chains of meat, milk, fruits and vegetables or wheat.

The Netchain model offers a way to illustrate the various vertical and horizontal transaction relationships in value chains of the agri-food industry during the production cycle. The interrelationships of the actors directly involved in production are in the foreground of this illustration. The relationships with and between supporting service providers of individual

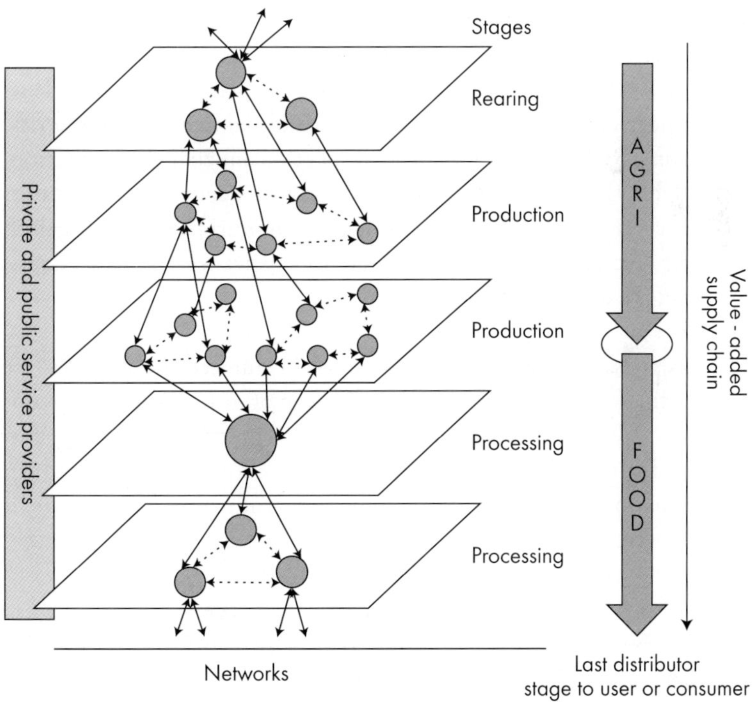

Figure 4.4. The Netchain model (modified from Lazzarini *et al.*, 2001; Schulze Althoff, 2006).

core actors (also across vertical stages) have been included additionally in the model by Schulze Althoff (2006).

Owing to the strong specialisation of the enterprises involved at all stages, different Netchains with varying customer-supplier relationships and contractual links exist in industry (Petersen, 2003; Spiller *et al.*, 2005). This heterogeneity regarding the many independent individual enterprises and the complex power structures within the sector, result in intercompany adaptation of quality and risk management activities presenting a special organisational challenge. The model depicted here helps to characterise the interrelationship of product and information flow of most of the supply chains to food retail or the catering trade and other last distributor stages to the consumer. The coupling of the two sectors, i.e. agri-food industry, via the 'bottleneck function' of the first processing stage can be seen as a key hallmark. This function is assumed for example by slaughterhouses, dairies, mills, breweries, fruit collecting and packaging points.

The AMOR model

The AMOR model represents a concept developed by Lang and Petersen (2012), where the formation of alliances between various actors of a food supply chain to jointly implement risk-oriented inspection strategies is in the foreground. The acronym AMOR stands for Alliances for the Mutual Organisation of Risk-oriented inspection strategies. This approach includes two perspectives, i.e. the collaboration for mutual benefit and the risk-oriented layout of inspections. Risk orientation in this respect means that the inspection intensity (what, when and how) depends on the probability of occurrence, the probability of detection and the severity of the effects for food safety, in short the risk. If the probability of occurrence and the severity of the effects are high, the inspection of the products is carried out more intensively than is the case with a low risk.

The AMOR-concept essentially has four advantages: (1) limited (human) resources can be better utilised in situations of high risk; (2) duplication of inspections of the same product batches at various stages can be avoided; (3) the development of new inspections in the chain can be carried out jointly, and thus the individual effort can be reduced; and (4) the exchange of information accompanying product and production between the individual stages can be improved.

Box 4.1. Inspection types.

- Online inspections: inspections taking placing within production.
- Offline inspections: inspections taking place outside production, e.g. by external laboratories.

The alliance of companies within the framework of an AMOR inspection takes a mutual decision about the planning and implementation as well as financing of an inspection strategy. This also includes how the organisational structure, risk-oriented inspection configuration and distribution of tasks and responsibilities of the partners are to be organised. The alliance has to develop an effective information and communication structure for successful introduction of the strategy. Cost/effort and benefits are distributed to achieve a win-win situation for all parties involved. The multitude of possible alliances can be subdivided into three main categories (Figure 4.5).

In category 1 the inspection is organised between companies having a direct customer-supplier relationship with each other. An example for this is the agreement between fruit and vegetable wholesalers and the producers where the mutual risk-oriented inspection is a joint pre-harvest inspection (as condition for harvest release) and a reduced incoming inspection at the delivery of the harvested batch. Another example is the decision of two alliance partners for a temperature-time indicator to control maintenance of the cold chain during product transport between supplier and customer.

In addition to customers and suppliers, an independent third party (private) entity may be part of the alliance. The third party acts as a mediator and can organise, coordinate and implement the inspections outlined in category 2. This third-party management of procedures requires the approval of all companies involved. It is important that the third party can also profit from the inspection. One example for this are the voluntary inspections to determine the health status of piglet producing enterprises from which fatteners buy their animals. The determination of the health status includes coordinated monitoring activities between the vendors and the buyers of piglets as well as registration and communication of the findings of the investigation organised by an animal health agency TiGA or by other netchain coordinators (O'Hagan *et al.*, 2013) (see Chapter 8.2).

Furthermore, inspections can be aligned and coordinated between customer, supplier and public or semi-public authority as an independent third party (category 3). Organisation

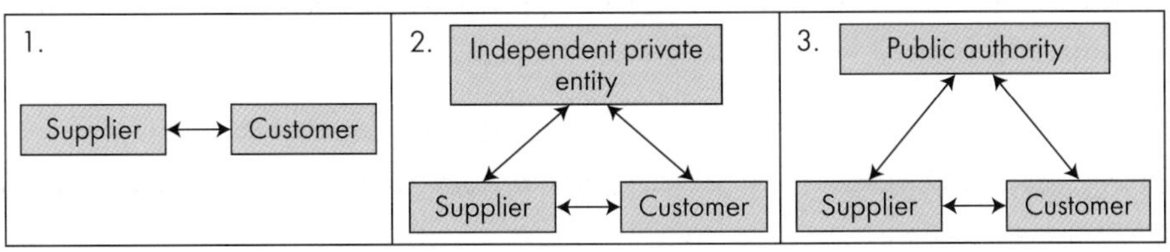

Figure 4.5. Three categories of alliances for the mutual organisation of risk-oriented inspection strategies (AMOR) (Lang and Petersen, 2012).

and realisation of the milk quality inspection in Germany via state control associations presents such an example. Another one is the *ante-* and *post-mortem* inspection where usually a governmental organisation (veterinary office) performs the inspection and will submit the results to the customer and supplier.

Figure 4.6 shows, on the one hand, the already illustrated sequence of incoming, intermediate and outgoing inspections, and, on the other hand, the combination of incoming and outgoing inspections of customer and supplier within the framework of the AMOR model.

At present, companies are mostly carrying out an incoming inspection at the point-of-arrival of the product. Then another outgoing inspection is performed prior to the transport of the sold goods. Normally, at least one intermediate inspection during handling, processing or storage is also carried out in the company. The information obtained during these inspections is mostly passed on to the next stage together with the product, rarely, however, information is returned to the previous stage. A flow of information in both directions is decisive for an improvement of product and process quality and to avoid a repetition of inspections as well as to advance the efficiency of the entire chain.

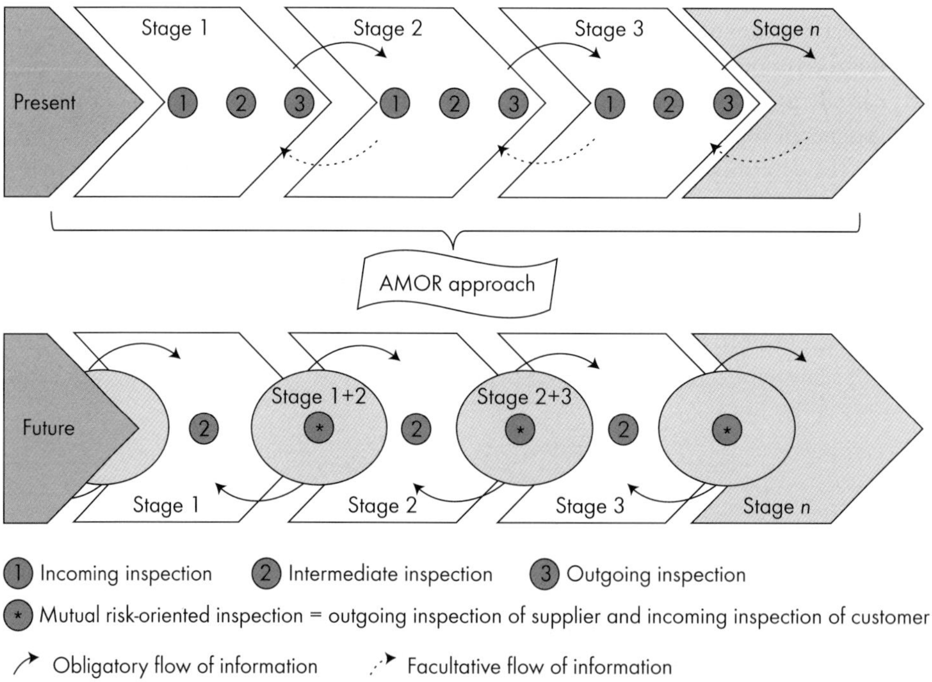

Figure 4.6. Effect of organising risk-oriented combined incoming and outgoing inspections in value chains (Lang and Petersen, 2012).

An example for a combination of the incoming inspection of the customer (stage 2) with the outgoing inspection of the supplier (stage 1) to achieve one joint inspection strategy is the organ diagnosis of slaughtered pigs at the slaughter line. Regarding the entire supply chain, not only the frequency of inspections could be reduced in this way, but also the number of inspection points. Information about safety and quality of the meat referred to the classification of the carcass halves and the organ diagnosis flow back to each animal owner as the supplier of slaughtered pigs who receives valuable information for farm management. At the same time, the slaughterhouse uses this data to control the process of cutting and processing. AMOR reflects a new trend in all sectors of the agri-food industry where more importance is attached to cooperation, transparency and quality communication.

The 3-level model

In the light of transposition of the EU Food Law (see Chapter 3), different scopes of duties have been newly organised in quality management. Within three levels of responsibility, at the moment normative, strategic and operative activities are developing at a fast rate to improve intercompany quality management (Figure 4.7). Within the framework of this development, new bodies and organisational structures have been established. Bodies composed of representatives from private and public organisations lay down rules, standards and laws at the normative level. Companies and those responsible at the strategic and operative level of production, processing and trade align their activities for quality assurance and quality management according to these requirements. Chapters 3 and 5 describe these requirements in detail.

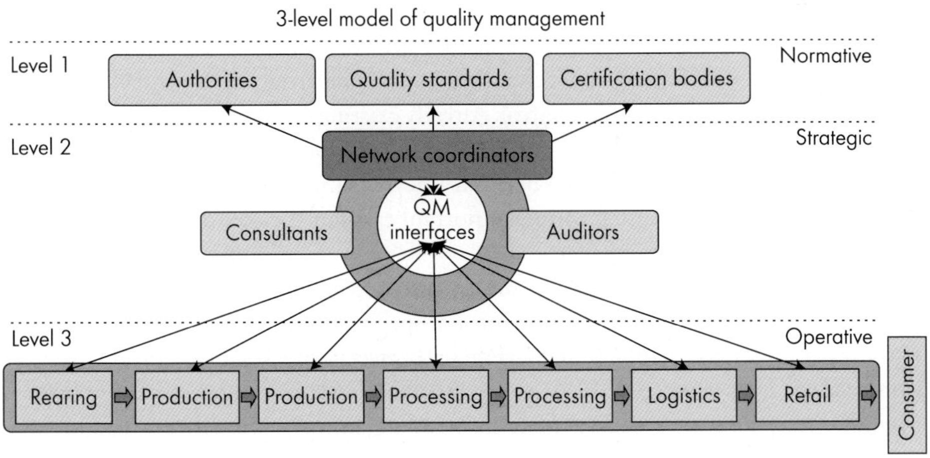

Figure 4.7. 3-level model (modified from Brinkmann *et al.*, 2011; Petersen *et al.*, 2007).

The actors of level 1 are thus defining the current framework for quality management coordinated throughout the industry. The food law, for example, prescribes efficient self-control systems according to the HACCP principles (see Chapter 2) for all stages of the value chain. This includes, amongst others, to fulfil the documentation duty. The voluntary exchange of information between customers and suppliers of the food chain supports traceability of delivery channels and is usually controlled by means of sector-specific standards. In addition to this, companies and associations of the agri-food industry and trade have decided in the past few years on an extensive catalogue of agreements in the form of inspection criteria within quality assurance programs (e.g. QS, IKB) and obligations to provide proof by supply chains within the framework of food standards (e.g. IFS, BRC) prescribed by trade (see Chapter 5.2).

First of all, many farmers considered the new legal provisions and new customer requirements merely as a burden. Only with time, did they become aware of the added value for their own quality management. Self-control and documentation requirements were considered as time-intensive and unreasonable at the stage of primary production (level 3). If, however, technical aids to reduce multiple recording are available and the data collected supports decision finding to improve product and process quality, the critical attitude often changes (Schulze Althoff, 2006).

The organisations of level 2 have a decisive relevance in this connection (Petersen *et al.*, 2007) since intercompany service organisations take over important coordination and control functions. At this strategic level, they support quality management of each individual company and the entire value chain in their function as 'bundlers' of many agricultural enterprises (see Chapter 3 and Chapter 8.2).

The urgent need for coordination developed with the introduction of the 'control of control' principle (see Chapter 3), i.e. establishment of industrial self-control systems while simultaneously restructuring food control. Already existing structures for the coordination of agricultural enterprises, such as producer associations or marketing cooperatives, have been strengthened, and where this was necessary, new organisational structures were created (Spiller *et al.*, 2005). Amongst the most important tasks of the coordinators at level 2 are the following:
- passing on and target-group specific processing of specifications from level 1;
- monitoring and auditing as well as control of specifications;
- coordination of the supplier management between the stages of production and processing;
- bundler functions in respect of trade in breeding animals, animals for slaughter, raw milk, cereals, fruits and vegetables;
- networking of consulting facilities (see Chapter 8.2);
- supporting farmers in their documentation tasks;

- supporting farmers in the passing on of information to downstream production stages and authorities;
- processing of process information according to demand from different systems of the individual production stages and allocating them to user groups.

One variant of the concept of a network coordinator is illustrated by Brinkmann *et al.* (2011) with the model of the quality broker. The broker function between the seller of a product and the purchaser relates to or is restricted to the coordination of transport activities, exchange of information required for safeguarding the business process, and pricing, including additional information such as the production conditions in an enterprise. Similar to a quality management system, the broker (Figure 4.8) takes over responsibility for selecting and assessing suppliers on behalf of the buyer or customer and determines customer satisfaction and carries out market analyses on behalf of the seller or supplier.

The models described so far (Netchain model, AMOR model, 3-level model) structure illustration of interrelationships in management and quality communication in the daily business of the actors involved in a Netchain. In exceptional situations and crises caused by animal diseases, natural catastrophes, management failure or criminal acts possibly

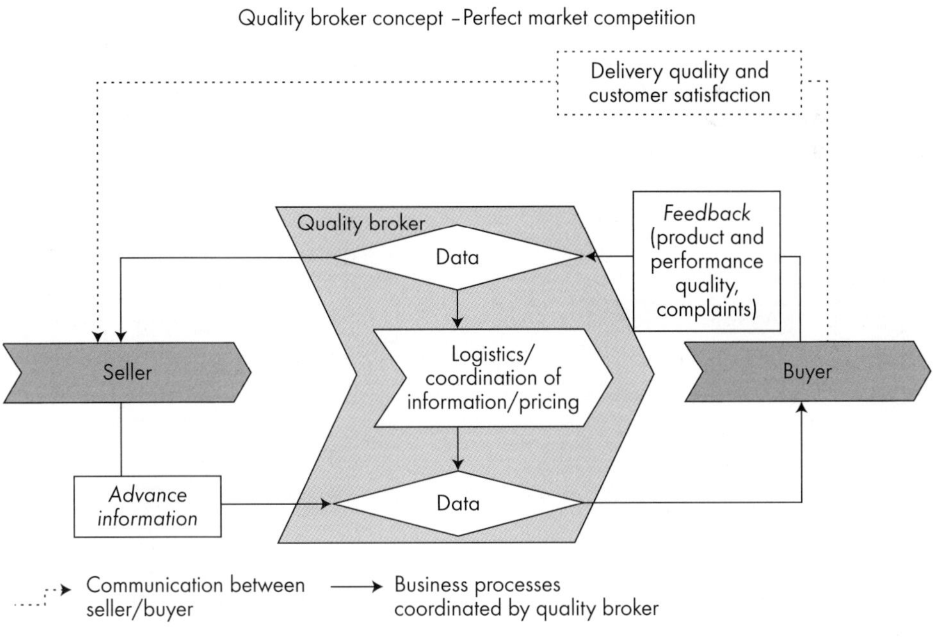

Figure 4.8. The concept of the quality broker as a variant of the 3-level model (Brinkmann *et al.*, 2011).

affecting individual stages or the entire value chain, the need for information as well as the willingness to communicate between the actors change all of a sudden. The concept and strategy of a controlled exchange of information in times of crisis to be realised in the long term are illustrated in the so-called engage-exchange model. The groups of actors presented here and the exchange of information amongst them are being checked for practicability at the moment (Petersen, 2012b; Schütz and Petersen, 2010; Slütter, 2013).

The engage-exchange model

The Engage-Exchange Model (EEM; Figure 4.9) is an approach to optimise crisis management. Public and also private information systems are equipped for short-term data exchange and additional collection and evaluation of data to support those responsible for crisis management in their decisions in a crisis situation.

The EEM distinguishes two different types of categories for data storage and processing, i.e. the 'active area' and the 'passive area'. Active areas are those areas required for quality communication during the on-going production process and during 'idle phases' (e.g. production data or performance data). Passive areas are possibilities for data storage and evaluation only provided for times of crisis (e.g. proof needed for traceability, consignment notes and receipts for medicines). These hardware and software solutions provided for

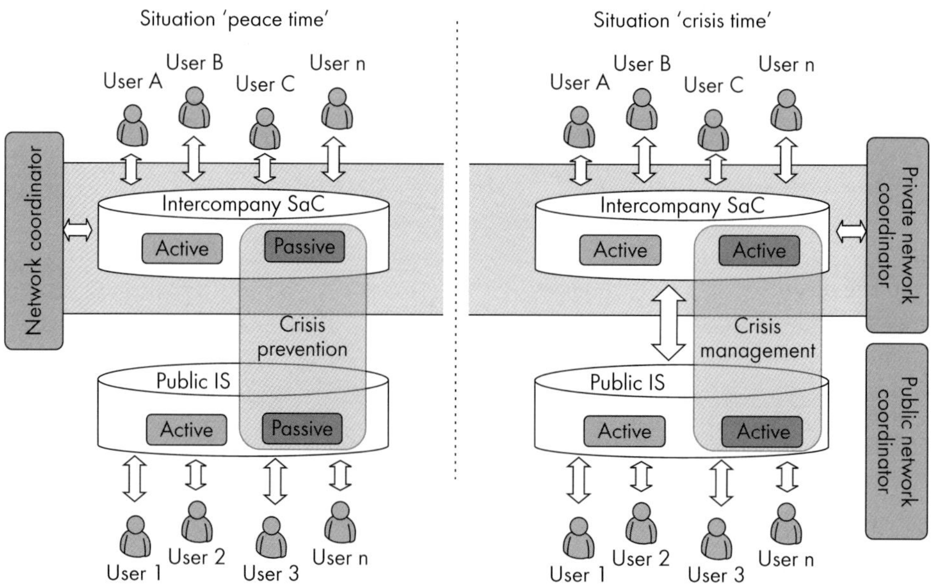

Figure 4.9. Engage-exchange model.

crisis prevention are to cover the specific information needs during a crisis. Data which needs to be supplemented for example when diseases occur, i.e. engaged, are additional information about contact persons, contact enterprises, etc. The quality codes and data collected routinely and during the process in connection with the engaged additional data serve in 'times of crisis' to manage the crisis (Breuer *et al.*, 2008; Schütz and Petersen, 2010). The technical and organisational infrastructure created for this, is part of crisis prevention.

Technical implementation of the EEM is done ideally by directly linking database systems or independent of systems via data portals. Furthermore, basic organisational conditions for realisation must be fulfilled. In this regard, network coordinators support organisational implementation by means of their bottleneck function within the supply chain. They can address a multitude of actors within the chain and thus become suitable multipliers for information. Network coordinators therefore have a key position in the provisioning of data (Brinkmann and Petersen, 2010; Schütz and Petersen, 2010). In particular, jointly coordinated guidelines defining the situation and the time for engaging and exchanging data and determining the responsibilities, competences and data exchange during the crisis, are important for organisational implementation (Slütter, 2013). In Chapter 9.2, possibilities of application of this model for crisis prevention and crisis management are explained in detail by way of examples.

The model of public-private risk management

The engage-exchange model illustrated in the previous section was used as an example to explain the importance and necessity of cooperation between private companies and public institutions for future-oriented decisions in crisis management. The federal institutions BfR (Federal Institute for Risk Assessment) and BVL (Federal Office of Consumer Protection and Food Safety) and their tasks in the field of risk analysis according to the EU Food Law are already known from Chapter 3. Risk management as generic term for the steps of risk identification, risk assessment, risk management, risk control and risk communication (Codex Alimentarius Commission, 1999) is to be considered as the task of the government regarding health-oriented consumer protection and also as an elementary component of company management (DIN ISO 31000). Corporate risk management above all deals with financial risks (Blanke, 2002). For the agri-food industry, integrated risk management including as well risks regarding food safety and food quality makes sense (Schmitz, 2006).

The possibility that products produced or processed by a food enterprise will not meet the requirements of food quality or food safety is definitely to be considered as a risk. This risk can lead to a loss of company image, market loss, loss of turnover, complaints, claims for compensation and liabilities (Schmitz, 2006).

The model of public-private risk management is an approach to explain how these two views can be combined. The purpose of the model is to find for the same terms in risk management definitions and procedures for the relevant typical roles and tasks of those responsible in public institutions and in companies or intercompany service organisations. The model wants to explain ideally and typically how private and public organisations in the agri-food industry deal with risks, and how these differing tasks and responsibilities and strategies are interlinked.

Layout of risk management with relevant control actions is measured by social objectives of health and consumer protection. General social and political protection objectives are the starting point for risk analysis and risk assessment by public institutions (e.g. Federal Institute for Risk Assessment (BfR)). When doing this, product-specific food safety objectives and general protection concepts (such as HACCP) are determined. De Swarte and Donker (2005) subdivide the path from public risk management to risk management carried out by individual enterprises in four phases:

1. Risk identification and risk assessment of foodstuff produced at home and imported.
2. Definition of general protection objectives and product-specific objectives to minimise risks.
3. Transformation of recommendations for private risk management (within the meaning of the EU Food Law).
4. Implementation of enterprise-specific HACCP concepts as the basis for risk assessment of concrete corporate processes.

Schmitz (2006) describes how companies of the agri-food industry integrate methods of risk assessment, such as HACCP and FMEA, into their own risk management. Food enterprises deduce from laws, regulations, standards, guidelines and recommendations their own intercompany-specific product and process specifications. Corporate risk management supports this transformation process:

- during product development;
- during improvement of (hygiene) processes; and
- by the development of an own concept of systematic risk assessment (Schmitz, 2006).

The public, food-law risk management (public) and the business risk management (private) are interconnected in several places (Figure 4.10):

- The quality policy of companies in the agri-food industry is essentially characterised by legal requirements regarding food safety and food quality.
- Consumers (predominantly indirectly via purchase behaviour and preferences) and the trade (frequently directly via private food standards) pose concrete to general requirements on the quality policy of food companies.
- Food companies transfer requirements (e.g. documentation duty, monitoring measures) from other areas (e.g. environmental law, policy of sustainability) to their own risk policy.

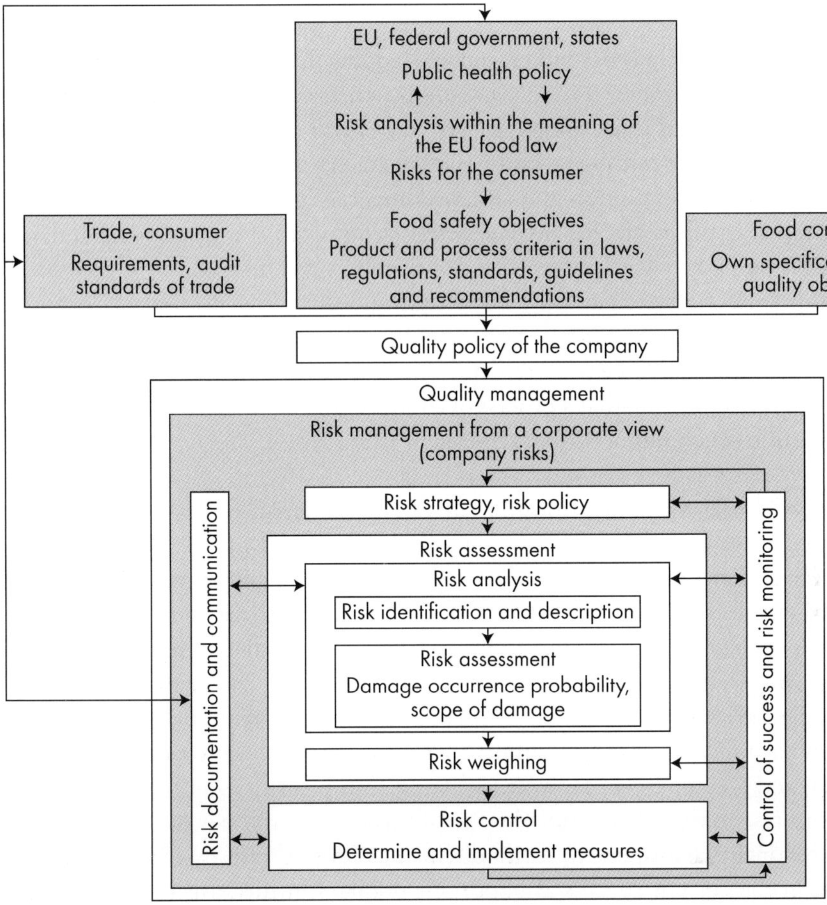

Figure 4.10. Combining risk analysis within the meaning of the EU Food Law with risk management from a corporate point of view (modified from DGQ, 1998; Hoornstra and Notermans, 2001; The Institute of Risk Management, 2002; Woidich, 2004).

- Risk management within the meaning of the food law provides for support of direct responsibility of companies.
- Risk communication by companies considers risk communication supported by authorities, and risk management by authorities, in turn, takes the interests of the agri-food industry into account.
- Successful consumer protection and health policies pursue within the meaning of the 'control of control' principle feedback from corporate risk control to risk analysis and to risk management of food safety authorities (see Chapter 8.1).

Companies have to define their risk policy, and when doing this, determine their company-specific measures to minimise risks dependent on their objectives. The first step of implementing risk-political objectives is to carry out a risk analysis. It is less important whether a nomenclature according to ISO 31000 characterised by commercial principles (risk identification, risk description and risk assessment) is used, or if the terms of food hygiene (hazard identification, hazard characterisation, exposure assessment and risk characterisation) are taken as the basis. It is decisive, however, to find out if this task is performed unprejudiced, systematically, completely and including all relevant corporate sectors. Table 4.1 contrasts various risk strategies of public institutions and companies or value chains of the agri-food industry. Public-private partnership alliances will play a special role in this respect making sure that a process of coordination of different risk cultures and different forms of risk communication take place.

Within the framework of risk identification, first of all, potential faults, their causes and resulting damages are determined and systematically illustrated in the risk description. Probability of occurrence of the risks and the scope of damage to be expected are categorised in damage classes and combined in a risk matrix (risk portfolio) to carry out qualitative risk assessment. Experience in using risk portfolios is explained in detail in Chapter 8.4.

An acceptability line subdivides the risks in acceptable and extreme risks. This enables in connection with risk weighting, alignment with risk-political acceptability limits, on the one hand, and a comparison between risks and thus prioritisation for planning the following measures, on the other hand. The aim of this must be to carry out an assessment with preventive quality management methods (such as FMEA) already when developing products and in the current process at important points. On the basis of the results of the analysis, it is possible to assess the effects of possible faults, interruptions, etc., which concrete risks are connected with them, and if measures are necessary or sufficient. These measures are implemented within the framework of risk control. In principle, four types of alternatives for action are available for this: prevention of risks, minimising of risks, passing on of risks and taking over the risk individually. Depending on the type of risk involved, certain forms of action are not available (e.g. taking over the risk individually or passing on the risk to insurance companies). It is important that the relevant path of risk control does not represent a final result, but necessarily a feedback to corporate risk assessment and corporate risk policy takes place to maintain continuous control of success and a continuous improvement process. Chapters 8 and 9 contain examples for this.

Table 4.1. Comparison of risk strategies in public and private organisations of the agri-food industry.

Risk strategy	Public institutions	PPP alliances	Industry
Risk policy	consumer and user-oriented protection of the consumer against health damages and fraud proper information for consumers	embedding a joint risk culture improvement of understanding of risk and risk communication	customer and supplier-oriented prevention of competitive disadvantages and image loss
Risk assessment	product and process-related measurement models for damage potentials determination of limit value	minimisation concept determination of signal values (see Chapter 8.2)	company or chain-specific FMEA or HACCP concept (Chapter 9) probability of occurrence significance + probability of discovery of risk factors, dangers and faults
Risk prevention	product-related monitoring risk knowledge platform principle of precaution = identification of potentially upcoming effects, assessment of available scientific data and scope of scientific uncertainty	PPP exercises for crisis management (Chapter 9) risk avoidance strategies	monitoring procedures AMOR principle (Chapter 4) passing on of risks (insurance companies)
Risk management	conversion of emergency and crisis plans in coordination with geographical and public responsibilities creation of task forces	application of engage-exchange models for risk and crisis communication	conversion of company or chain-specific emergency and crisis plans

AMOR = alliances for the mutual organisation of risk-oriented inspection strategies; FMEA = failure mode and effects analysis; HACCP = hazard analysis and critical control points; PPP = public-private-partnership.

4.2 Quality control cycles

B. Petersen
[1]University of Bonn, Katzenburgweg 7-9, 53115 Bonn, Germany
b-petersen@uni-bonn.de

Each quality management system requires a control cycle registering, checking and assessing the QM measures to prescribe, if need be, corrective measures. A plan-do-check-act cycle represents such a control cycle (Figure 4.11). DIN EN ISO 9000 (see Chapter 5.2) also

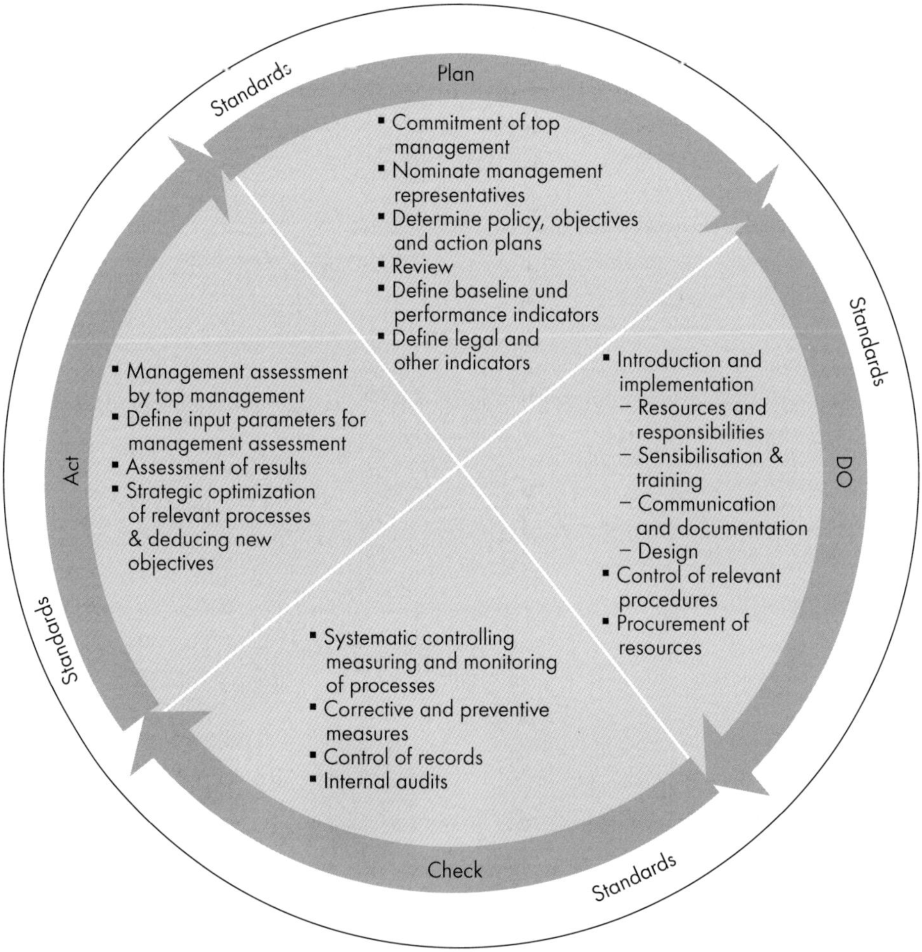

Figure 4.11. Plan-do-check-act cycle according to different standards.

describes a control cycle as well as the EFQM model (see Chapter 3.1) and a series of further standards for systemising and illustrating management tasks:

- The phase of *planning* summarises all obligations of the top management for implementing the quality policy.
- In the *implementation* phase (do), processes are determined and personal and financial resources provided to achieve the quality objectives.
- In the next step, systematic controlling and verifying of all processes follows (*check*).
- Finally, regular *assessment* of the quality management system and a target-actual comparison between target figures determined, and actual figures, which could be achieved, leads to a new cycle being run through. On the basis of the difference between quality objective and measurable actual parameters, the quality policy and the action plans for a new period are defined anew.

The overall objective is to initiate a continuous improvement process and run through it time and time again within an organisation. The role of the people assuming responsibility for this and taking decisions is in the foreground here.

The control cycle with feedforward control

This particular role perception in management can be illustrated even better by means of the control cycle model (Figure 4.12). A control cycle is technically speaking the closed control circuit which continuously registers the variable to be controlled, compares it with the standard variable (target variable) and is influenced in such a manner as to bring about alignment with the target value.

We speak of a quality control cycle if the variable to be controlled is a quality characteristic (Petersen, 1985; Strotmann, 1989). The socio-technical consideration takes into account that the controller (e.g. QM representative, management, board of directors and board of supervisors) as well as the controlled system (e.g. individual company, group of companies) include human beings as elements. The principle of the quality control cycle is therefore also part of non-technical processes, i.e. in inter-departmental or intercompany control cycles.

A fundamental distinction between different control cycles is made by grouping them into internal and external quality control cycles. If, for example, a customer is influencing the production process of a supplier by his quality request, two information and control cycles are interlinked, one of which is an implementation process and the other the decision process running outside the organisation. The default values necessary for control can be determined in both cases internally and externally (for example by customers, suppliers, laws, etc.).

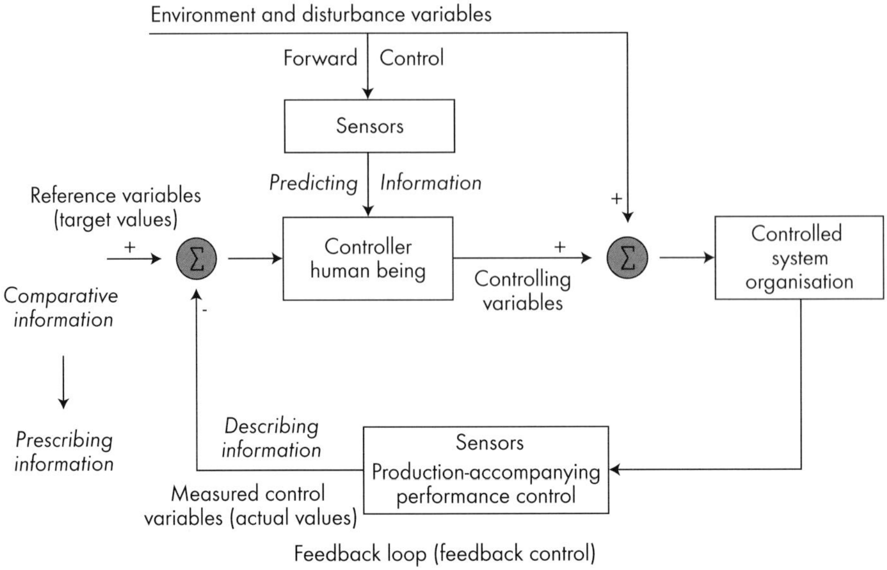

Figure 4.12. Schematic diagram of a socio-technical control cycle model with feedforward control (modified from Petersen, 1985; Strotmann, 1989).

The controller, and thus the human being, is taking intuitive as well as considered decisions. For the latter, four categories of information are important: descriptive, comparative, predicting and prescribing information. The task to determine and implement time and time again independent quality objectives adapted to changing framework conditions is embedded in the understanding of the quality manager as the controller. He is thus taking over management and performance tasks in the socio-technical system, in an individual company or across companies referred to in cybernetics as open and closed-loop control tasks.

In contrast to the principle of closed-loop control, open-loop control always includes feedback to monitor the measures undertaken. With an open-loop control, the management can react better to short-term changes in the production process and takes its decisions not rigidly and according to certain formulas as would be the case with a closed-loop control system. If the management notices deviations from the target or setpoint value early on, it can introduce corrections by targeted action.

In a control cycle model with feedforward control, the most important known disturbance variables are measured directly and are compensated by control. This enables to keep the effects on the controlled system, i.e. the organisation, as small as possible. The feedback cycle (control) thus presents the link between open and closed-loop control.

The control cycle of decision processes

The essential steps of the decision process – from recognising the problem, selecting the right alternative of action up to finally taking a decision – can be illustrated as a control cycle model. Figure 4.13 shows the phases run through conscientiously but partly also unconsciously.

In conclusion of the functionally essential of the open and closed-loop control processes described, one can say that quality management of an organisation involves influencing a system by procuring, processing and passing on of information from the system and its subsystems.

The controller of the system, in this case the quality manager, uses this information not only in a result-oriented manner but also in an action-oriented way. Planning and control represent important management tasks in this respect. They can be illustrated as shown in Figure 4.13 as closely interconnected phases of information processing. As a result, quality and risk management are understood to be continuously running processes of decision

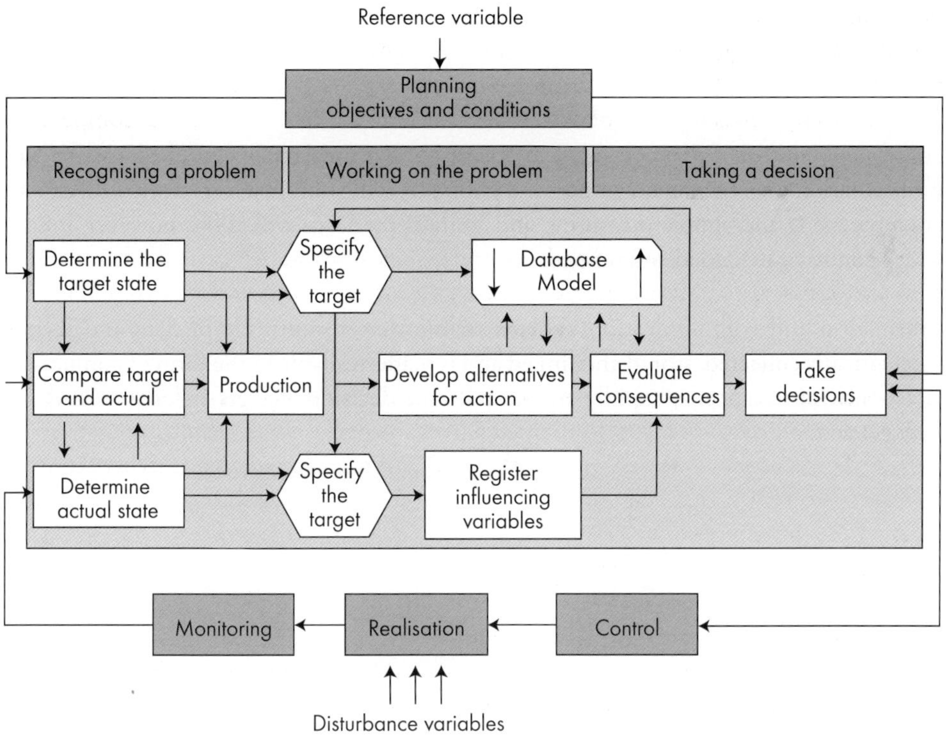

Figure 4.13. Control cycle of decision process (Petersen, 1985).

finding. The management takes its decisions considering the conditions specified for the organisation. The quality objectives simultaneously represent target values which are later compared with the actual production result to obtain comparative information or 'control information'. Each decision process thus implies collection and processing of information.

The control information therefore has a special importance since it first enables to identify problems and to analyse them. If the problems have been recognised, the target-actual comparison provides information to demonstrate which influential variables or actions could have caused that the planned target was not reached, but also information about optimum alternatives of action. This requires, however, that in addition to the 'describing' and 'comparing' information, 'predicting' and 'prescribing' information are available as well.

Prescribing information are answers to the question 'What will happen, if...?', e.g. 'What will happen if a deviating quality is only detected by the customer and not in the company?' Prescribing information are details about optimum alternatives for action, e.g. 'increase the frequency of tests!', and go back to the initial situations where an 'if-then-relationship' can be established.

First of all, it is, however, necessary to regularly collect raw data during production. They form the basis for processing (compression) to achieve the categories of information mentioned (Figure 4.14).

Continuous control of all processes of an organisation is a real challenge. The human power of observation and combination necessary for this will soon reach its limits. Electronic data processing and simple measuring and analysis methods will allow, however, human perception to be extended and refined.

Information and communication systems enable that a multitude of data and facts is systematically collected, stored and linked and is made available in the form of compressed data – i.e. information – rapidly to the relevant decision makers in quality and risk management.

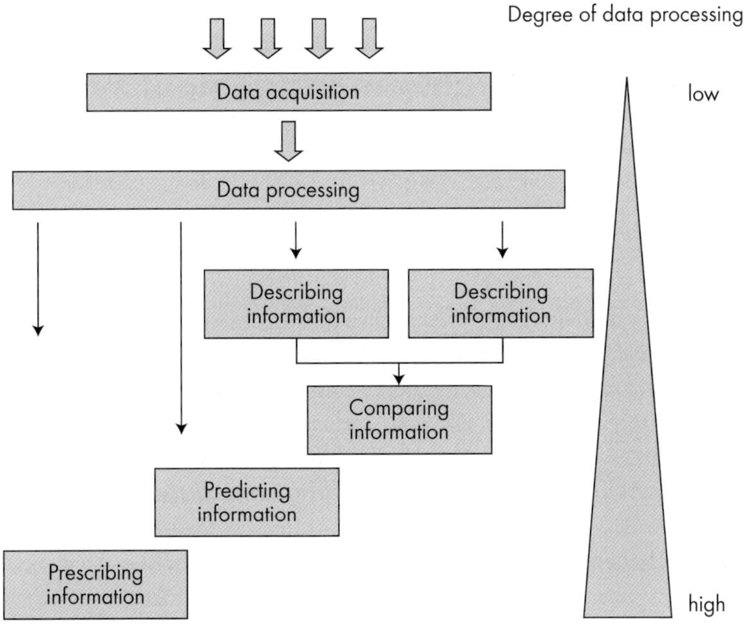

Figure 4.14. Levels of information increase by data processing (Ellebrecht, 2008).

4.3 Procedural models for the design of intercompany quality management systems

V. Raab[1] and B. Petersen[2]
[1]DRV, Deutscher Raiffeisenverband e.V., Pariser Platz 3, 10117 Berlin, Germany
[2]University of Bonn, Katzenburgweg 7-9, 53115 Bonn, Germany
raab@drv.raiffeisen.de

Intercompany quality management systems have to be planned and systematically designed in the long term. Several research projects have developed procedural models and pilot-tested their application. From these, general concepts and strategies for intercompany quality management systems in the agri-food industry can be deduced.

The stage concept for realising traceability and quality communication

One of the major challenges when designing a quality management system is efficient and target group-oriented configuration of the exchange of information between the individual stages of a value chain (Berns, 1996; Horvath, 2004; Petersen, 1985; Petersen *et al.*, 2002; Pfeiffer, 2005; Poignée *et al.*, 2005; Schiefer and Helbig, 1995). Special drivers for this are the necessity to adhere to laws and consider standard requirements, for example within the framework of identification and traceability of products and documents.

A comprehensive procedural model for designing and developing intercompany information systems is the stage concept according to Schulze Althoff (2006). He defines 6 consecutive stages from planning up to introduction and use of a data warehousing system providing the actors of a supply chain to the food retail with quality-relevant information. He determines the following objectives:

1. Generating a generic pattern of testing and decision criteria to determine which quality information has to be provided.
2. Integrating methods of statistical process control and controlling for visualisation of disturbance variables.
3. Long-term processing and storing of data by correspondingly integrating technologies in the solution approach.

The stage concept defined by Schulze Althoff (2006) (Figure 4.15) is a guideline representing a methodical and systematic framework for the introduction and operation of systems to ensure distributed tasks in intercompany quality management. The concept not only demonstrates possible ways to proceed to those responsible in companies in the private sector but also to authorities. Two stages refer to the procedure during planning and preparing intercompany information systems. The next four stages describe the steps necessary for its introduction and use.

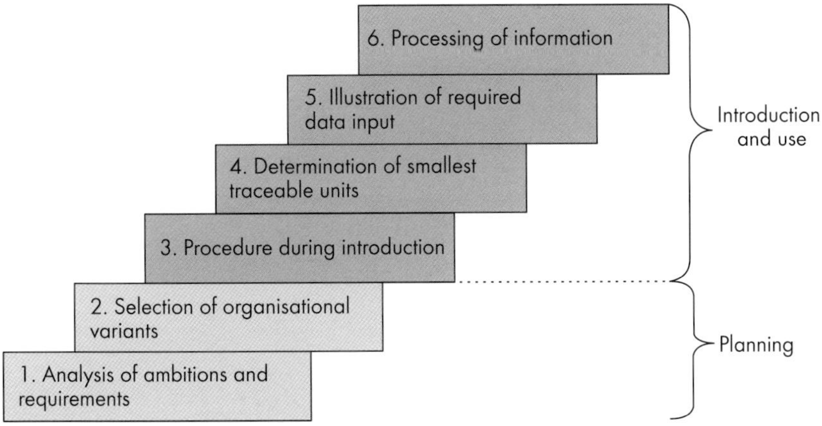

Figure 4.15. Stage concept for the conception and implementation of intercompany information and communication systems (Schulze Althoff, 2006).

In the first stage, the ambitions of the potential system users are identified. An illustration of the typical requirements of the relevant user groups allows interested parties when designing their own information systems to match them with the demands of their own target group. On the basis of the second stage, different user groups are enabled to design their information management system according to the relevant form of organisation of the actors in the value chain or Netchain (see Chapter 8.2). Stage 3 describes the technical conceptual design to optimally identify the interfaces between intercompany core and supporting processes. Determination of the smallest traceable unit is based on an agreement between customer and supplier of a Netchain and thus determines the degree of detail and intensity of forward and backward information. It is also essential to determine the degree of data exchange and the number and rights of user groups whose forward and backward information is to be organised (according to the control cycle models illustrated in Chapter 4.2). The sixth step in the concept includes processing of information by means of business intelligence applications. The type and scope of these analyses depend on the objective and concrete issues of the individual user groups.

The TOF strategy – evaluation of improvement potentials

The systematic approach to simultaneously view technical (T), organisational (O) and functional (F) aspects when planning an information and communication system is also described as TOF strategy (Figure 4.16). Several authors point out that this three-dimensional view is required to actually achieve a continuous improvement process in all companies involved in a value chain with an intercompany ICT system (Beulens, 2003; Petersen *et al.*, 2007; Raab, 2011; Schulze Althoff, 2004; Schulze Althoff *et al.*, 2005; Trienekens, 1999). Quite often, rapid introduction of intercompany ICT systems fails due

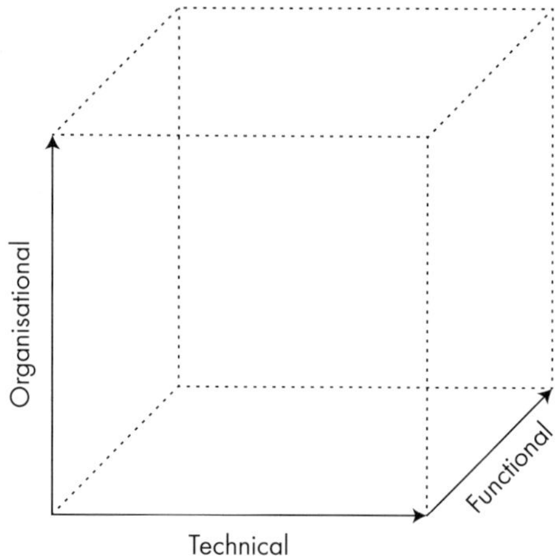

Figure 4.16. Schematic diagram of the three-dimensional view for continuous improvement (Schulze Althoff, 2006; Raab, 2011).

to the fact that only the new technical means are considered and the change processes in the organisational area and the decision-relevant audit functions to determine data are initially not taken into account.

The technical dimension refers in this context to the use of already tried and tested technologies, integration of relevant data from existing data sources and the inclusion of established ICT systems. Regarding the organisational dimension of assessment and planning, several authors suggest to consider aspects such as responsibilities in intercompany management, confidence amongst actors, market power and position and readiness for cooperation when designing intercompany information management systems. The functional dimension of assessment includes identification of specific instruments and tools enabling determination of data. These might be, for example, online and offline inspections or audits providing raw data.

Raab (2011) and Bruckner *et al.* (2013) developed exemplary for the area of cold-chain management a set of assessment tools including these three mentioned dimensions. One of the tools is called the 'multi-criteria decision aid – cold chain management (MCDA-CCM)' concept. This tool enables systematic planning of the introduction of intercompany temperature monitoring systems. By means of objective criteria, different technologies for temperature measurement and data evaluation are assessed. At the same time, the relevant potentials for improvement of cold-chain management are illustrated and the different measuring systems ranked. This is necessary since the improvement process of cold-chain management inside the company and across several companies requires not only considering technical aspects, the functional and organisational aspects have to be taken into account as well. The MCDA-CCM tool supports safe investment decision taking in cold-chain management. This is due to the fact that in every chain the best alternatives for cold-chain management can be determined by means of a calculation program. This approach includes the outranking method promethee (Preference Ranking Organisation Method for Enrichment Evaluation). It compares several solutions even if their technologies differ greatly, and only small preferences for a certain solution exist.

The MCDA-CCM concept (Figure 4.17) consists of the two steps 'analysis of requirements relevant for value chains' and 'assessment and decision for a system solution'. Each phase is subdivided further. In the first step, use of the 'three-dimensional generic model' provides for concrete determination of technical, organisational and functional criteria for the inspection technologies to be assessed in temperature mapping. Numerical weighting of 14 criteria determined in the 'scoring model', forms the basis for subsequent assessment steps.

Concept phases	1. Analysis phase		2. Evaluation and decision phase	
Steps	1st step Requirement analysis in the cold chain	2nd step Evaluation of technical, organisational and functional alternatives	3rd step Determination of calculation variables	4th step Ranking of alternative solutions
MADM model level	1st model level		2nd model level	
Method / model	Three-dimensional generic model (TOF)	Scoring model	MCDA method (PROMETHEE)	
Process elements of CCM	Analysis of improvement potential in cold-chain management	Calibration of 14 relevant criteria	Determination of weighting coefficients for different temperature-controlled products	Ranking of temperature monitoring systems
	Verification of relevant criteria	Evaluation of temperature monitoring alternatives	Threshold values for the durability of different products	

Figure 4.17. Matrix structure of the assessment concept (multi-criteria decision aid – cold chain management) for evaluating and ranking of temperature monitoring possibilities (Raab, 2011). CCM = cold chain management; MADM = Multi Attribute Decision Making; MCDA = multi-criteria decision aid – cold chain management; PROMETHEE = Preference Ranking Organisation Method for Enrichment Evaluation; TOF = technical, organisational and functional.

On the basis of this, in the second phase, the multi-criteria decision method promethee (Brans *et al.*, 1986; Rousis *et al.*, 2008; Schütz, 2009) is used. The available calculation tool enables calculation of ranking values in the third step. In the fourth step, finally, ranking of preferred solutions for different application scenarios to improve the relevant management in food producing chains is carried out.

Procedural model for assessing the benefit of ICT in intercompany quality management

Can the benefits accompanying the introduction of an ICT system be quantified? Ellebrecht (2008) describes an interesting methodical approach to answer this question. The author suggests calculating a specific benefit index for each user group of an intercompany ICT system as a common investment for a supply chain to food retail.

By means of the index formula developed, dimensionless codes can be calculated for each individual actor and system functionality. They permit ranking the benefit which the individual actors derive from the use of internet-based information and communication systems. From this, he can deduce a key for a fair distribution of costs for joint investments. The model can be used in different situations, such as before taking the decision to establish an intercompany information and communication system or when planning to extend an already existing solution.

The benefit indices calculated for each specific target group are suitable for carrying out a detailed cost and benefit assessment taking the cost for implementation and extension of information and communication systems into account.

Ellebrecht (2008) determines five parameters to calculate the benefit index according to the following formula:

$$\text{Benefit index} = \frac{ZG(1+\Delta d) \times IZ \times \ddot{U}G}{ZE(d)}$$

where:
IZ = information gain;
ZG = time gain in days;
$\ddot{U}G$ = degree of similarity in data exchange;
ZE = period for decisions in days (d).

The dimension *time gain*, ZG (*in days*) describes the difference between the time needed before and after introduction of the information and communication system for processing and provision of user-specific information.

Where ZG (time gain in days): $1+\Delta d$ = Alternative I in d – Alternative II in d (Alternative I: time needed for providing information without using the intercompany information and communication system. Alternative II: time needed for providing information after introduction of the intercompany information and communication system).

The range of variation of the parameter results from the minimum to maximum possible time gain for the observation period (e.g. one month of use). When calculating the benefit indices for each actor, the observation period has to be selected equally for each system functionality and user.

When defining the parameter 'information gain' (IZ), a distinction is made between several levels according to the intensity of data processing and the resulting information categories. The information gain arises from reaching a certain level with a higher degree of processing. Index calculation always takes the highest possible degree of processing into account. IZ (information gain) is categorised into the following when doing this (according to the illustration of the corresponding control cycle of information, Chapter 4.2):

- 1 = digitised data;
- 2 = describing information;
- 3 = comparing information;
- 4 = predicting information;
- 5 = prescribing information.

The variation range of the parameter is between 1 and 5.

The parameter 'degree of similarity in data exchange', ÜG (dimensionless) can be calculated according to (Ellebrecht, 2008) as follows from three dimensions: ÜG = GB × GIR × GNE (explanation of the dimensions: degree of willingness to exchange information, GB (I-VI), degree of traceability, GIR (I-IV), degree of expected benefit, GNE (I-III)). By multiplying these three dimensions, a variation range of the parameter from 1 to 72 results.

'Period for decisions', ZE (in days) is defined as the timeframe available to the actors in health management of individual companies or across companies to intervene in production or consulting processes in a controlling or regulating manner (e.g. piglet production, fattening, stock supervision, procurement of animals). The range of variation of the parameter results from the minimum (>0) to maximum possible period for decisions for the observation period (e.g. month of use).

The basic principle of this procedural model of systemising a benefit observation was taken up and further developed by other authors (Düsseldorf *et al.*, 2011; Schütz, 2009). In Chapter 8.2 an application example for this is described.

Procedural model for planning service offers in intercompany quality management

As already illustrated in the three-level model, network coordinators support intercompany quality management with a series of service offers. Schütz (2009) suggests a procedure involving several steps to plan these systematically. When doing this, she takes both sides into account, i.e. the potential service users and the service providers.

She works with the hypothesis that the service quality is best when it is possible to close all gaps in the communication between service user and service provider. For the GAP model, Zeithaml *et al.* (1990; Figure 4.18) define quality as the degree of discrepancy between expectations or requirements of the customers and their perception of the actually provided service. Deviations correspond to the differences between the service expected by the customer and the service perceived by the customer. From the GAP model, six sequential work steps to measure service quality and thus customer satisfaction can be deduced:

- determination of customer expectations;
- specification of the expected services;
- analysis of the service provision process;
- specification of the services to be provided;
- verification of communication of service provision potentials;
- assessment of service quality by the customer.

How services can be planned concretely is explained by Schütz (2009) in the so-called AAD (Analysis, Assessment, Decision) model (Figure 4.19). It describes a systematic build-up of three phases (analysis, assessment, decision) and seven steps to select the right service offer for a customer or a customer group.

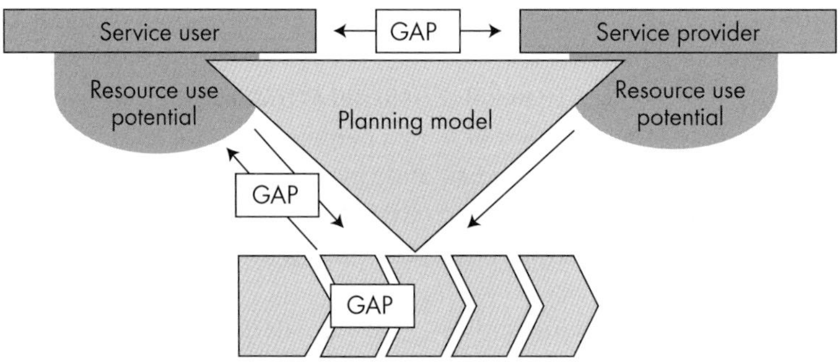

Figure 4.18. GAP model (Schütz, 2009).

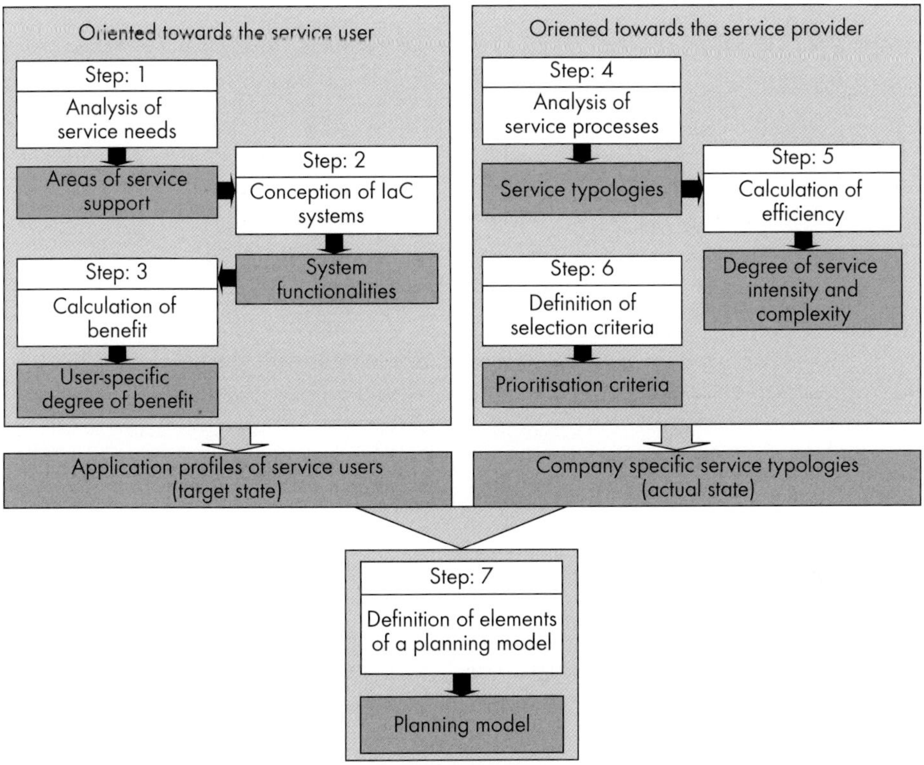

Figure 4.19. The seven steps of the AAD planning model (Schütz, 2009).

She bases this approach on four assumptions:

- Network actors involved in the value process 'pork' in primary production need sector-specific communication structures to guarantee quality and safety of the end product better.
- The decisions of different actors in quality management of individual companies or across companies can be supported in a demand oriented manner by information and communication systems.
- An optimum combination of suitable service typologies for specific groups of service providers can be planned on the basis of measurable criteria.
- Strategic planning is a process in 7 steps.

First of all, the variables influencing service quality are analysed, then a benefit observation for information and communication systems between service providers and service users is carried out and ultimately this procedure is verified when introducing new services.

Chapter 5. Certification procedures and foundations

Quality is an absolute must. Objective audits or certifications ensure the reliability of products, procedures, services or systems regarding their quality and safety. At the same time, they guarantee conformity with the requirements of corresponding standards, guidelines and laws.

This chapter will inform you about:
- the applicable standards and labels in the agri-food industry;
- the structure of the German accreditation scene;
- the specific certification models applying for the agri-food industry.

Keywords: international standards, labels, certification, harmonisation, audit, certification models

Summary

In the agri-food industry, DIN EN ISO 9001 represents the core structure for establishing management systems. Nevertheless, industry and sector-specific standards determine the certification landscape. This includes standards to assure process and product quality of foodstuff and feedstuff such as the QS certification seal most wide-spread in the value chain of meat, fruits, vegetables and eggs. Conformity tests are therefore carried out increasingly within the framework of combined audits. For agricultural enterprises special certification models have been developed: group certification and quality association certification. Accredited industry-specific certification bodies have proven that they carry out their work competently adhering to legal and normative requirements and at an internationally comparable level. In this regard, certification measures have to be seen, first of all, as measures to build confidence in the customer-supplier relationship of the value chains while labels are rather used as instruments of quality communication towards the consumer.

5.1 International standards

S. Bruckner[1], S. Lehnert[2] and L. Theuvsen[3]
[1]AGRIZERT, Zertifizierungs GmbH, Vorgebirgsstr. 80, 53119 Bonn, Germany
[2]University of Bonn, Katzenburgweg 7-9, 53115 Bonn, Germany
[3]Georg-August-University Göttingen, Department of Agricultural Economics and Rural Development, Platz der Göttinger Sieben 5, 37073 Göttingen, Germany
sbruckner@agrizert.de

Standardisation is the process by which a standard is developed. In Germany, this process is governed by the DIN 820 series which has been harmonised with the Internal Regulations of CEN, the European Committee for Standardisation, and the Directives of ISO, the International Organisation for Standardisation. DIN EN 45020 defines Standardisation as the '...activity of establishing provisions for common and repeated use with regard to actual or potential problems aimed at the achievement of the optimum degree of order in a given context' and states in the following that 'important benefits of standardisation are the improvement of the suitability of products, processes and services for their intended purposes, prevention of barriers to trade and facilitation of technological cooperation'. The interested parties mentioned in this definition include, amongst others, manufacturers, trade, commerce, consumers and service providers, academia, test institutes and the state. The German Institute for Standardisation (DIN) more or less provides the 'round table' where the representatives of the interested parties meet to carry out standardisation work. This is defined by the following principles: voluntary support of standardisation, publicity, participation of all interested parties, consensus, relevance, international character and with an orientation towards the latest scientific and technical findings, the economic given facts and the general benefit.

The term standard is defined in DIN EN 45020 (DIN EN 45020, 2007-03) as a 'document established by consensus and approved by a recognised body that provides for common and repeated use rules, guidelines or characteristics for activities or their results (...).' Standards have no legally binding force but are rather recommendations whose application occurs on a voluntary basis. The most important standards in the agri-food industry are the IFS Food (International Featured Standard Food), QS (Qualität und Sicherheit GmbH), BRC (British Retail Standard) and GlobalGAP which are explained in more detail later on (Chapter 5.2). In the area of certification, standards are often referred to as 'certification systems'.

In connection with the standards used in Germany, quite often you will find the following abbreviations indicating if a standard is valid nationally, in Europe or internationally:
- DIN: German standard (Deutsches Institut für Normung e.V.);
- EN: European standard (established by CEN: Comité Européen de Normalisation);
- ISO: International standard (International Organisation for Standardisation).

The figure behind the colon of a standard designation indicates the year of the standard version, e.g. 2008 for DIN EN ISO 9001:2008. Table 5.1 contains an overview of common standards in the agri-food industry in the area of quality management and related disciplines which are illustrated in this chapter in more detail. Further relevant standards on the subject of quality management are listed in Table 5.2.

Almost all of the standards for implementation of the management systems presented here are based on the so-called PDCA cycle: Plan-Do-Check-Act, also designated Deming Circle (Figure 5.1). This circle was already introduced (Chapter 4.2) and can be described according to DIN EN ISO 9001:2008 as follows:

- *Plan:* Identifying the objectives and processes necessary for achieving results in accordance with customer requirements and the policies of the relevant organisation.
- *Do:* Implementing the processes.
- *Check:* Monitoring and measuring of processes and products by means of policies, objectives and requirements on the product and reporting on the findings.
- *Act:* Taking measures for continuous improvement of process performance.

Table 5.1. Standards in the field of quality management and in related disciplines.

Number of standard[1]	Title	Can be certified
DIN EN ISO 9000, 2005	Quality management systems - fundamentals and vocabulary	
DIN EN ISO 9001, 2008	Quality management systems - requirements	yes
DIN EN ISO 9004, 2009	Managing for the sustained success of an organization - a quality management approach	
DIN EN ISO 19011, 2011	Guidelines for auditing management systems	
DIN EN ISO 22000, 2005	Food safety management systems - requirements for any organization in the food chain	yes
DIN EN ISO 50001, 2011	Energy management systems - requirements with guidance for use	yes
DIN EN ISO 14001, 2009	Environmental management systems - requirements with guidance for use	yes
DIN ISO 26000, 2010	Guidance on social responsibility	
DS 49001, 2011	Social responsibility management system - requirements	yes
DIN ISO 31000, 2009	Risk management - principles and guidelines	
ONR 49001, 2010	Risk management for organisations and systems - risk management - Practical use of ISO/DIS 31000	yes

[1] DIN: Deutsches Institut für Normung (German Institute for Standardisation); EN: Europäische Norm (European Standard); ISO: International Organisation for Standardisation; DS: Dansk Standard (Danish Standard); ON: Österreichisches Normungsinstitut (Austrian Institute for Standardisation; ONR: ON-Regel (ON rule).

Table 5.2. Overview of relevant standards in the field of quality management (Source: National Annex to DIN EN ISO 9001, 2006).

Number of standard	Title
DIN EN ISO 10012	Measurement management systems - requirements for measurement processes and measurement equipment
DIN ISO 10001	Quality management - customer satisfaction - guidelines for codes of conduct for organisations
DIN ISO 10002	Quality management - customer satisfaction - guidelines for complaints handling in organisations
DIN ISO 10003	Quality management - customer satisfaction - guidelines for dispute resolution external to organisations
DIN ISO 10005	Quality management - customer satisfaction - guidelines for quality plans
DIN ISO 10007	Quality management - customer satisfaction - guidelines for configuration management
DIN ISO 21747	Statistical methods - process performance and capability statistics for measured quality characteristics

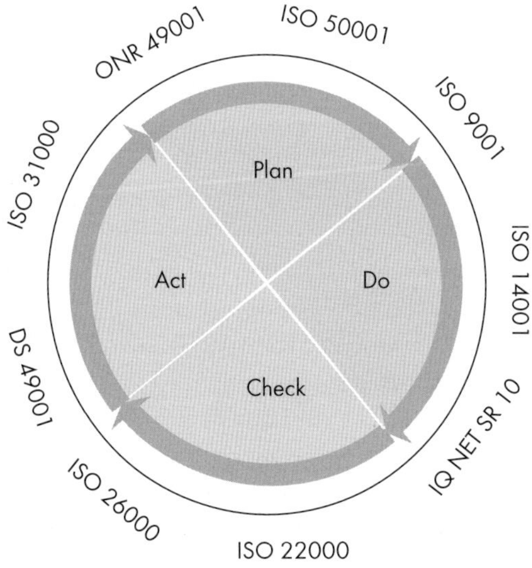

Figure 5.1. The PDCA (Plan-Do-Check-Act) cycle in international standards.

Orientation of standards along the PDCA cycle also facilitates integration of various management systems in an already existing management system to design a so-called integrated management system (IMS) (see Chapter 2.2).

Quality management

The family of standards DIN EN ISO 9000 ff. represents fundamental international and intercompany standards in the field of quality management. Since its revision in 2000, there only exist the following standards in their current version; the standards DIN EN ISO 9002 and DIN EN ISO 9003 have been withdrawn at that time:

- DIN EN ISO 9000:2005: quality management systems – fundamentals and vocabulary;
- DIN EN ISO 9001:2008: quality management systems – requirements;
- DIN EN ISO 9004:2009: managing for the sustained success of an organisation – a quality management approach.

Only DIN EN ISO 9001 is a standard for certification.

DIN EN ISO 9000 describes fundamental aspects of quality management systems (QMS) and defines eight management principles on which the 9000 family of standards is based:

1. *Customer focus.* Organisations depend on their customers and should therefore understand current and future customer needs, should meet customer requirements and strive to exceed customer expectations.
2. *Leadership.* Leaders establish unity of purpose and direction of the organisation. They should create and maintain the internal environment in which people can become fully involved in achieving the organisation's objectives.
3. *Involvement of people.* People at all levels are the essence of an organisation and their full involvement enables their abilities to be used for the organisation's benefit.
4. *Process approach.* A desired result is achieved more efficiently when activities and related resources are managed as a process.
5. *System approach to management.* Identifying, understanding and managing interrelated processes as a system contributes to the organisation's effectiveness and efficiency in achieving its objectives.
6. *Continual improvement.* Continual improvement of the organisation's overall performance should be a permanent objective of the organisation.
7. *Factual approach to decision making.* Effective decisions are based on the analysis of data and information.
8. *Mutually beneficial supplier relationships.* An organisation and its suppliers are interdependent, and a mutually beneficial relationship enhances the ability of both to create value.

Furthermore, DIN EN ISO 9000 defines relevant terms in the following fields:

- quality-related terms;
- management-related terms;
- organisation-related terms;
- process and product-related terms;

- characteristics-related terms;
- conformity-related terms;
- documentation-related terms;
- investigation-related terms;
- audit-related terms;
- terms related to quality management for measuring processes.

DIN EN ISO 9001 focuses on the process-oriented approach for the development, implementation and improvement of an efficient QMS. The objective is to increase customer satisfaction by fulfilling customer requirements. For this purpose, the processes of the company are subdivided into four large areas forming the basis of every process-oriented QMS and representing the main chapters of the standard at the same time:

- *Management responsibility.* Management commitment, customer focus, quality policy, planning, quality objectives, responsibility, authority and communication, management review (Chapter 5).
- *Resource management.* Provision of resources, human resources, infrastructure, work environment (Chapter 6).
- *Product realisation.* Planning of product realisation, customer-related processes, development, purchasing, production and service provision, control of monitoring and measuring equipment (Chapter 7).
- *Measurement, analysis and improvement.* Monitoring and measuring, control of non-conforming products, data analysis, improvement (Chapter 8).

The PDCA cycle mentioned already at the beginning of this chapter can be applied to all four areas. The model of a process-oriented QMS according to DIN EN ISO 9001 including a description of the interrelationships of the various processes is illustrated in Figure 4.1.

As already mentioned, DIN EN ISO 9001 is the only standard of the 9000 family of standards which can be certified. When selecting an accredited certification body for certification according to DIN EN ISO 9001, approval of the certifier for the own industrial sector has to be considered. This so-called scope is indicated with 39 code numbers altogether and was issued by the IAF (International Accreditation Forum) and the EA (European Co-operation for Accreditation). This list of scopes of accreditation is based on the statistical nomenclature for economic activities NACE Rev. 2 published by the Commission of European Communities (EU, 2012b).

'Agriculture and forestry and fishing' is registered for example as Scope 1, 'Food products, beverages and tobacco' as Scope 3 and 'Other services' as Scope 35. The accreditation process and the interrelationship of certification and accreditation are detailed more closely in Chapter 5.3.

DIN EN ISO 9004 was developed in 2000 together with DIN EN ISO 9001 as a consistent pair of standards supplementing each other. It complements the requirements of DIN EN ISO 9001 with notes and recommendations. This is still the case after the revision carried out in 2009, whereas now the focus is on the sustainable success of an organisation. DIN EN ISO 9004 offers guidelines to the management of any organisation aiming at achieving sustainable success by a quality management approach within a complex, demanding and consistently changing environment. It has to be seen as a support for self-assessment and thus an important element of advancement and not as a guideline to apply DIN EN ISO 9001.

DIN EN ISO 19011 (Guidelines for auditing management systems) has to be seen as closely related with the 9000 family of standards. It describes management and control of an audit program, the procedure for carrying out audits and determining competence and assessment of auditors (Chapter 5.3).

Management of food safety

Since 2005, DIN EN ISO 22000 defines the requirements placed on a management system for food safety for organisations along the entire food chain as an international standard. This standard can be applied for organisations directly involved in the food chain (e.g. feed manufacturers, agro-businesses, food manufacturers, canteens, food retailers) as well as for those organisations involved indirectly in the food chain (e.g. suppliers of devices, packaging materials, cleaning agents and disinfectants). It combines the principles of the HACCP system (Hazard Analysis and Critical Control Points, see Chapter 2.1) with so-called preventive programs (PRPs). The hazard analysis is in the focus and aims at identifying, assessing and controlling all (micro)biological, chemical and physical hazards to be expected when the consumer is consuming the foodstuffs. The Codex Alimentarius, a collection of international food standards, decided on by a joint body (Codex Alimentarius Commission) of the Food and Agriculture Organisation (FAO) and the World Health Organisation (WHO) of the United Nations, recommends application of the HACCP concept already since 1993. The HACCP concept consists as a basic preventive quality management method according to the Codex Alimentarius Commission (2003) of seven principles describing application of the HACCP system in a company:

1. conduct a hazard analysis;
2. determine the critical control points (CCP);
3. establish critical limits for each CCP;
4. establish a system to monitor control of each CCP;
5. establish the corrective action to be taken;
6. verification of the HACCP system;
7. documentation.

In contrast to DIN EN ISO 9001, DIN EN ISO 22000 exclusively refers, however, to a management system for food safety and not to a quality management system. Yet, it is compatible with DIN EN ISO 9001, i.e. it has a similar structure and text. The essential chapters of DIN EN ISO 22000 are the following:
- *General requirements and documentation requirements on a food safety management system* (Chapter 4).
- *Management responsibility.* Food safety policy, planning, responsibility/authorities, communication, emergency preparedness and response, management review (Chapter 5).
- *Resource Management.* Provision of resources, human resources, infrastructure, work environment (Chapter 6).
- *Planning and realisation of safe products.* Prerequisite programs (PRPs), hazard analysis, HACCP plan, updating of information and documents, verification planning, traceability, control of non-conformities (Chapter 7).
- *Validation, verification and improvement of the food safety management system.* Validation of combined control measures, control of monitoring and measuring, verification and improvement of the management system (Chapter 8).

The appendices A-C of this standard contain cross-references to DIN EN ISO 9001 and HACCP and further references to the Codex Alimentarius documents.

Energy management

In many organisations and companies, energy consumption is amongst the most important cost factors. Therefore it is of essential importance to continually optimise energy consumption by means of an energy management system (EnMS). This system can help to reduce energy consumption and thus the cost involved systematically, comprehensively and in the long term. The requirements placed on an organisation regarding introduction, implementation, maintenance and improvement of an EnMS is described since 2012 by the certifiable international standard DIN EN ISO 50001 (Energy management systems – requirements with application guidelines) which superseded the European standard DIN EN 16001 from the year 2009. The concept of the standard is illustrated in Figure 5.2.

This standard is also based on the already known PDCA cycle which is reflected in the chapters of this standard as well:
- plan: energy policy and planning (Standard Chapter 4.3 and 4.4);
- do: introduction and implementation (Standard Chapter 4.5);
- check: verification (Standard Chapter 4.6);
- act: management review (Standard Chapter 4.7).

An EnMS according to DIN EN ISO 50001 can be implemented as an independent management system or combined within the framework of an integrated management

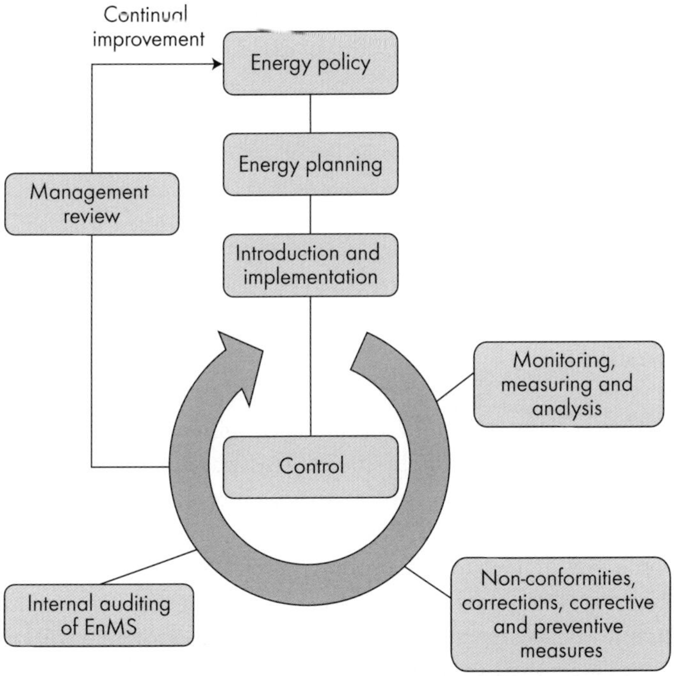

Figure 5.2. The concept of DIN EN ISO 50001 (DIN EN ISO 50001:2011).

system, e.g. with management systems according to DIN EN ISO 9001 or DS 49001. In particular, a combination with an environmental management system according to DIN EN ISO 14001 is recommended.

Environmental management

In addition to the EMAS II Regulation of the EC, the international standard DIN EN ISO 14001 is the most important requirement for environmental management systems in a company. This is, just like the family of standards DIN EN ISO 9000, also a family of standards. In addition to DIN EN ISO 14001 defining the requirements placed on an environmental management system as a certifiable standard, there are almost a dozen of other standards in this family of standards. Analogously to the DIN EN ISO 9000 family, for example DIN EN ISO 14004 represents a guideline for implementing DIN EN ISO 14001 (DIN EN ISO 14004, 2010: environmental management systems – general guidelines on principles, systems and support techniques). DIN EN ISO 14001 is based on the familiar PDCA cycle aiming at continuously improving environmental protection in a company by establishing, maintaining and constantly improving the environmental management system's efficiency. This standard is designed compatibly with DIN EN ISO

9001. The current version dates from 2009, in 2012, however, revision measures were carried out to create DIN EN ISO 14001:

> *Key aspects of the standard that are to be revised include a consideration of the entire value/supply chain and not only of the organisation itself, which was previously the case. The needs of small and medium-sized enterprises are also to be addressed to a greater extent. Special attention will be paid to enhanced user-friendliness and a simple, easily understood wording. (DIN EN ISO 14001, 2012).*

Social responsibility

In the past few years, sustainability and social responsibility have increased in importance in companies (Theuvsen and Friedrich, 2012). This was already recognised in 2004 in connection with ISO and resulted in an international standard being created which was ultimately published in 2010. DIN EN ISO 26000 is a guideline on social responsibility to be applied on a voluntary basis ('Corporate Social Responsibility', CSR). Not only industrial organisations but organisations of any kind – irrespective of their field of activity, size, owner structure, social context, culture or religious background – will be supported so that they are able to meet their social responsibilities better.

This standard defines seven principles, seven core subjects and 37 fields of action altogether. The following principles are defined as the basis and precondition for credible confrontation with social responsibility to be adhered to by any organisation: accountability, transparency, ethical behaviour, respect for stakeholder interests, respect for the rule of law, respect for international norms of behaviour, respect for human rights. Furthermore, organisations should deal with the core subjects organisational governance, fair operating practices, human rights, consumer issues, labour practices, community involvement and development and the environment (Figure 5.3). For each of these core subjects, Chapter 8 of the standard lists various fields of action for which corresponding expected actions are formulated.

Box 5.1. DIN ISO 26000 (DIN ISO 26000, 2010).

DIN ISO 26000 defines social responsibility as 'Responsibility of an organisation for the consequences of its decisions and activities for the society and environment by transparent and ethical behaviour which:

- contributes to *sustainable development*, including health and public welfare;
- considers the expectations of *stakeholders*;
- adheres to applicable *law* and is in accordance with *international standards of behaviour*;
- is *integrated* in the entire organisation and is lived in its relations'.

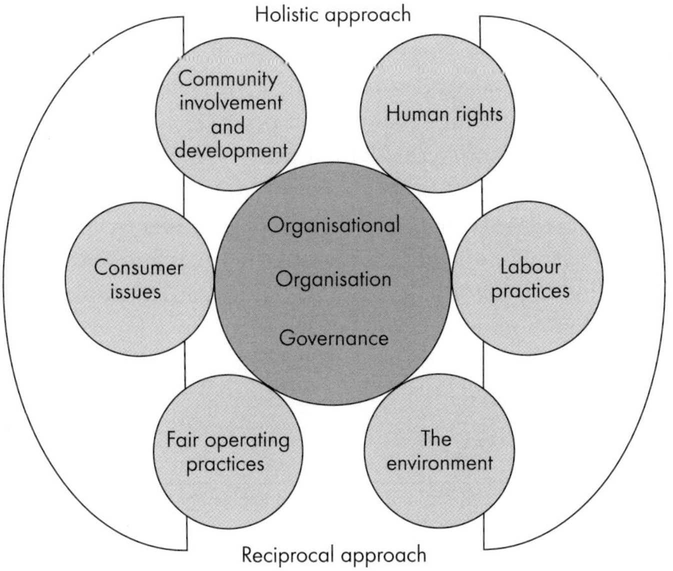

Figure 5.3. The seven core subjects of corporate social responsibility according to ISO 26000.

A detailed explanation of the principles, core elements and fields of action can be found in a publication of the Federal Ministry of Labour and Social Affairs (BMAS) (DIN ISO 26000, 2010) which is free of charge and published online, and on the homepage created especially by the BMAS on the subject of CSR[1] in Germany.

DIN ISO 26000 is a non-certifiable standard which is clearly emphasised in the standard:

> *This international standard serves for user orientation; it is not intended for certification purposes and also not suitable for such purposes. It would be a misinterpretation of the intention and purpose of this international standard to offer certifications according to DIN ISO 26000 or to make the claim that a certification according to DIN ISO 26000 exists (www.bmas.de/DE/Service/Publikationen/a395-csr-din-26000.html).*

The central organisations of the German Industry, BDA, BDI, ZDH and DIHK together with the involved governmental departments have formulated a common opinion clarifying that ISO 26000 is not to serve as a reference framework for certifications.

ISO 26000 thus defines no requirements at all for practical implementation of the principles of social responsibility. Concrete requirements and measures are increasingly sought after

[1] http://www.csr-in-deutschland.de.

by companies and organisations, however. Therefore, in 2010, the Association for Social Responsibility at Dansk Standard developed the certifiable standard DS 49001 'Social responsibility management system – requirement description'. This standard has the status of a national standard. Since January 2013, there exists an English translation of the Danish standard (DS 49001, 2012). It is also based on the already mentioned PDCA cycle, and regarding the basic structure of a management system, is similar to DIN EN ISO 9001 and DIN EN ISO 14001 (environmental management). The requirements and contents are based on the guideline ISO 26000. Accordingly, DS 49001 demonstrates sustainability and social responsibility in seven core areas which were also defined as the core subjects of DIN ISO 26000 (Figure 5.4).

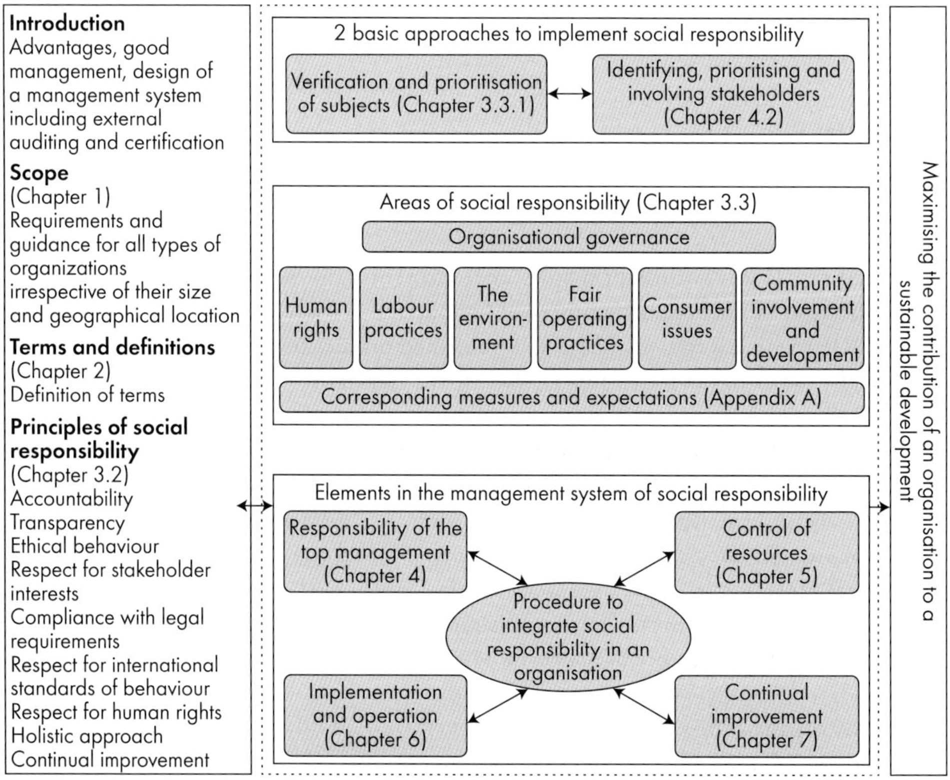

Figure 5.4. Overview of DS 49001 (DS 49001, 2011).

The seven core subjects are described by Rühl[2] as follows:

- *Organisational governance* means communication with stakeholders, i.e. the different interest groups of the company, transparent selection of suppliers and applicants and comprehensive risk management.
- *Human rights* is subdivided into two categories: on the one hand, adhering to civil and political rights and equality before the law, and on the other hand, paying attention to social, economic and cultural rights. But also the right to education, work and food plays an important role in this respect. Companies are to include, above all, groups in need of protection to avoid discrimination and exclusion. In times of globalisation, this is a central subject for most of the companies.
- *Labour practices* contains a legal overview including the issues of hiring employees, termination of employment contracts and remuneration. This subject also contains issues of health and safety at the workplace and personnel development. Companies are to maintain a regular dialog between authorities and other state institutions and their employees.
- Responsible handling of *the environment* is also a subject of the standard. Companies must encourage clean technologies and environmental awareness and therefore handle resources responsibly. Accident scenarios should also be prevented, and handling of substances dangerous for human beings and the environment should be weighed up considerably.
- In the field of *operating practices*, fair behaviour towards business partners, suppliers and employees is of importance: this includes, above all, avoiding corruption and corruptibility. Furthermore, the competition and advertising law is to be respected and company information protected.
- It is important for companies to look after *consumer issues* in order to improve customer relationships: confidential handling of customer data, against the background of data protection, has to be given as much priority as well as organised complaint management and a competent and well informed service.
- Companies are part of the society, and therefore the last topic *community involvement and development* is of utmost importance: companies and organisations should act in any matters economically responsibly and make their knowledge and technologies available to the public, but at the same time be also engaged in the areas of education and cultural activities and support them actively.

DS 49001 has an integrative character. A management system based on this standard can be integrated into other already existing management systems (e.g. a quality management system according to DIN EN ISO 9001) or serve as the basis for an integrated management system.

---

[2] http://www.business-netz.com/Qualitaetsmanagement/Norm-DS-49001-misst-Nachhaltigkeit-und-gesellschaftliche-Verantwortung.

In addition to the Danish standard DS 49001, almost at the same time (December 2011), the international certifiable specification *IQNet SR 10*, developed by the IQNet Association (an international certification network) and based on the RS10 specification designed by AENOR in Spain, was published. The IQNet SR 10 specification determines requirements for definition, implementation, maintenance and improvement of a management system for social responsibility to integrate social responsibility in all organisations and be able to contribute to a sustainable development taking the needs and expectations of stakeholders into account. Just like DS 49001, this specification is based on the PDCA cycle and the contents of DIN EN ISO 26000. It is compatible with other management systems (e.g. DIN EN ISO 9001) and applicable to all types of organisations. At the moment, this standard is only available in English and can be requested via the homepage of the IQNet Association (www.iqnet-ltd.com). DS 49001 as well as the IQ-Net SR 10 specification ultimately pursue the objective of converting the contents of ISO 26000 as a guideline to be applied on a voluntary basis into a document with certifiable requirements.

Risk management

The standard ISO 31000 was in 2009 the first globally valid standard on the subject of risk management. It contains terms and definitions, principles of risk management, a description of the elements of a risk management system and the risk management process. Three specific principles are embodied in this standard: risk management is presented as a management task comprising a comprehensive top-down approach and a generally applicable basic standard. ISO 31000 is also oriented towards the PDCA cycle with the risk management process being part of the systemic approach according to Brühwiler (2009) (Figure 5.5).

Certification according to ISO 31000 is not possible. Instead of this, the Austrian standard ONR 49001 can be used which already in its subtitle deals with practical conversion of ISO 31000 (ONR 49001 'Risk management for organisations and systems – Risk management – conversion of ISO 31000 into practical use'). It was, amongst others, the basis for creating ISO 31000 and is basically similar to this standard, its objectives and contents go, however, beyond: the elements of the risk management system are illustrated in a differentiated manner to ensure that it can be verified objectively. ONR 49001 belongs just like DIN EN ISO 9001 or DIN EN ISO 14001 to a family of standards illustrated in Figure 5.6. The certifiable standard ONR 49001 is supplemented by standards including three guidelines, terms and principles and the requirements on the qualification of the risk manager.

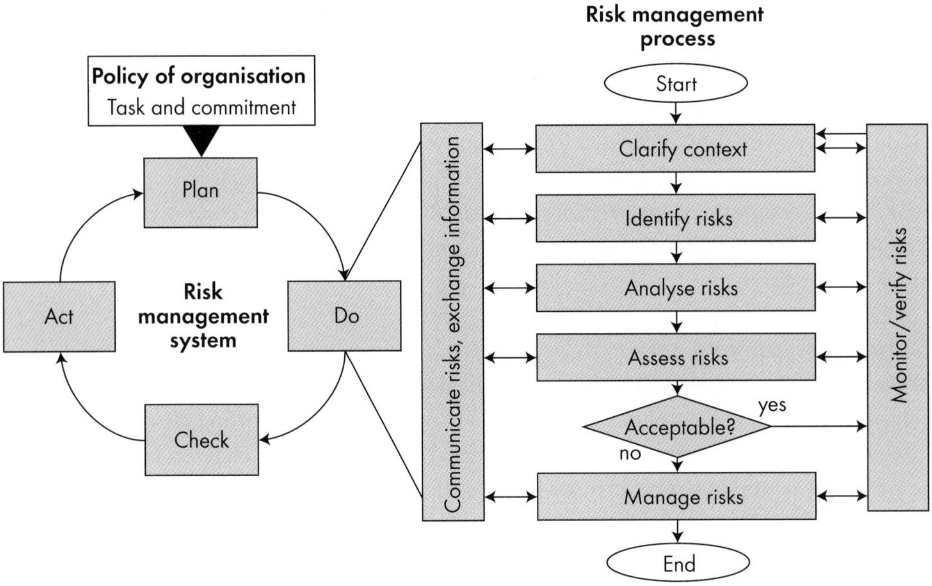

Figure 5.5. PDCA (Plan-Do-Check-Act) cycle of risk management (Brühwiler, 2009).

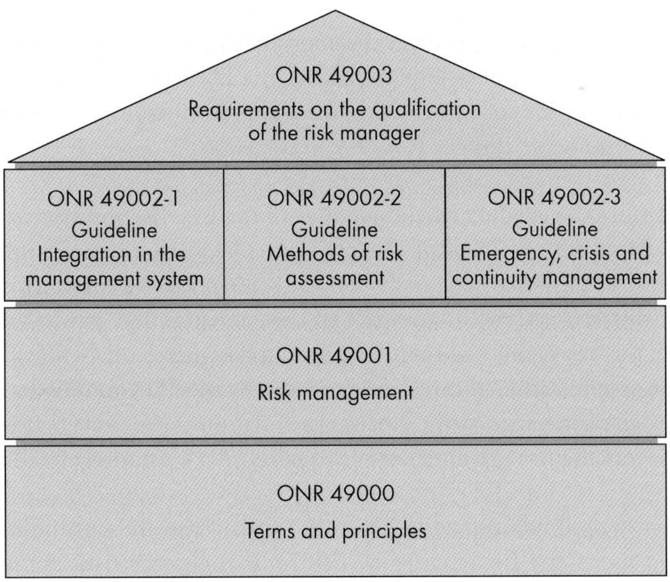

Figure 5.6. Risk management for organisations and systems (ONR 49001, 2010).

5.2 Labels and certification systems

L. Theuvsen[1], S. Bruckner[2] and S. Lehnert[3]

[1]Georg-August-University Göttingen, Department of Agricultural Economics and Rural Development, Platz der Göttinger Sieben 5, 37073 Göttingen, Germany

[2]AGRIZERT, Zertifizierungs GmbH, Vorgebirgsstr. 80, 53119 Bonn, Germany

[3]University of Bonn, Katzenburgweg 7-9, 53115 Bonn, Germany

ltheuvs@gwdg.de

Foodstuff has manifold quality characteristics (Luning and Marcelis, 2009). These are in part search properties whose particularities the end consumers can already determine before their purchase (e.g. colour of the meat), but partly also experience properties which can only be assessed after consumption (e.g. taste). A third group of quality characteristics finally has the nature of confidence properties whose particularities cannot even be verified after the purchase of the food or only with prohibitively high cost (e.g. no residues or no use of gene technology, origin, animal welfare standards, organic versus conventional production). Experience and confidence features are the foundation of information asymmetries which can be exploited by opportunistic producers and suppliers to their benefit. The consequences of information asymmetries can be a failure of the market (Akerlof, 1970) or – since foodstuffs are so-called 'necessary goods' with a very inelastic demand (Hardes and Uhly, 2007) – financial and even health damage to the consumers. Therefore, major efforts have always been made to protect the consumer against fraud by incorrect product labels (Grüne, 1994). This applies in particular to the EU which pursues the objective of 'labelling which will provide consumers with necessary information to enable them to make safe, healthy and sustainable choices' (DG SANCO, 2006).

Figure 5.7 systemises the product labelling policy of the EU. The approaches followed can be distinguished, on the one hand, if the labels document that basic requirements have been fulfilled or the fulfilment of quality beyond this regarding product or cultivation characteristics (differentiation). In addition to this, a distinction can be made, between certification, on the one hand, and rules for product designation (labelling), on the other hand. The scope of application of certification is seen by the EU Commission rather in the area of complex characteristics, while labelling regulations have proven effective when it comes to simple statements regarding product quality (EU Commission, 2009).

The certification systems mentioned in Figure 5.7 today have an outstanding importance in the agri-food industry (Theuvsen *et al.*, 2007). It is characteristic for a certification system that neutral external controls (audits) are carried out by independent, accredited parties (certifiers) (so-called third-party audits, Figure 5.10) leading to certification of a company (Meuwissen *et al.*, 2003). The basis for this is a certification standard formulated by a standard setter. The companies are audited on a regular basis and – if necessary – in

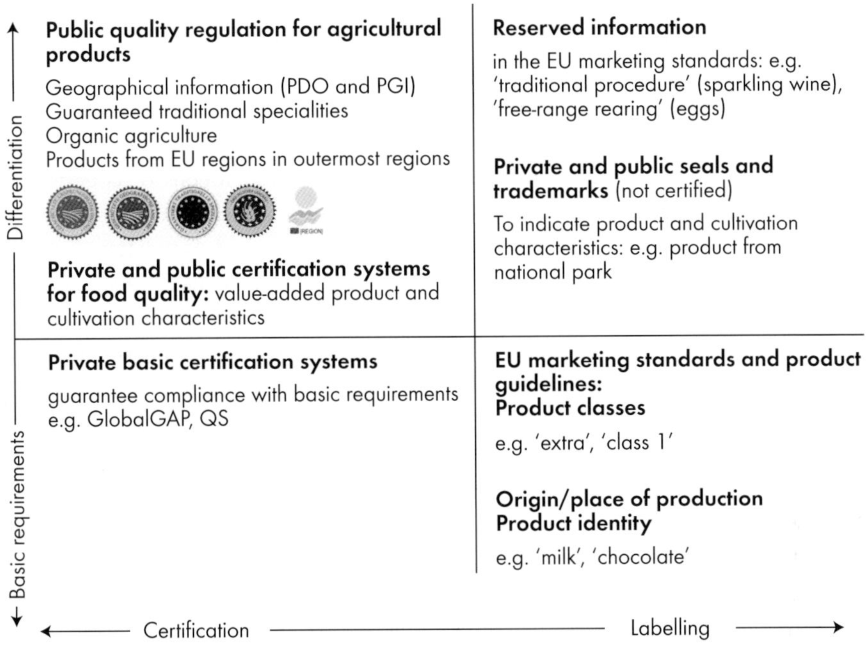

Figure 5.7. Systematics of the product labelling policy of the EU (based on EU Commission 2009: 5).

addition to this from case to case or randomly. The certificate awarded to the controlled company serves as a quality signal for purchasers or consumers. This is particularly relevant in markets characterised for example by purchase relevance of confidence properties and thus by quality uncertainties.

There exist a great number of certification systems; now, the overall number in the EU corresponds to approximately 360. The multitude of systems can be systemised according to different criteria. The most important are the following (Gawron and Theuvsen, 2010):

- *Standard setter.* Coarsely, a distinction can be made between public and private certification standards (Jahn *et al.*, 2003). When it comes to public standards, the EU (e.g. organic certification according to the EC basic regulation on organic production No. 834/2007; EU, 2007), the German federal government so far not active in this area and the state governments (e.g. Basis-Qualitätsmanagementsystem (BQM, basic quality management system) Sachsen-Anhalt; Geprüfte Qualität (Tested Quality) – (GQ-) Bayern) could function as holders. Private standards are supported by purchasers (e.g. IFS Food), suppliers (e.g. Red Tractor Assurance for Farms in Great Britain), Certifiers (e.g. Geprüfte Qualität (Tested Quality) of the TÜV SÜD) or social groups (e.g. church organisations in case of GEPA/Fairtrade). In addition to this, combinations are also

Box 5.2. The QS scheme.

The German QS scheme is a certification system established in 2001 as a reaction to the BSE crisis for meat and meat products; since 2004, fruits, vegetables and potatoes can also be certified according to QS. QS Qualität und Sicherheit GmbH based in Bonn is supported by the Deutsche Raiffeisenverband e.V., Deutsche Bauernverband e.V., Verband der Fleischwirtschaft e.V., Bundesverband der Deutschen Fleischwarenindustrie e.V. and Handelsvereinigung für Marktwirtschaft e.V. Thus representatives from various value chains along the meat-producing chain participate in the QS scheme – true to its inter-stage approach. In 2004, the QS Fachgesellschaft Geflügel GmbH and the QS-Fachgesellschaft Obst, Gemüse, Kartoffeln GmbH were founded. QS thus is a private certification system initiated and supported by industry.

The criteria to be fulfilled by the participating companies within the framework of the QS scheme are defined for the various value chain stages in corresponding system handbooks forming the basis for external audits. If the audit has been successful, the QS certification mark can be used. Central objectives are, amongst others, to assure food safety by process quality, to offer more reliability to the market and stabilise consumer confidence. The QS scheme does not serve to differentiate products but to ensure minimum qualities in the market. It can be used in the B2B and also together with the QS certification mark in the B2C area. QS nowadays has more than 130,000 scheme partners, of them more than 11,000 abroad (status: 01.10.2011), and is thus one of the biggest certification systems in the food area worldwide. In Germany, in 2011 about 95% of the pork and poultry meat and about 70% of the beef production was QS certified (see www.q-s.info).

possible, for example representatives of different value chains establishing jointly a standard, for instance the QS scheme.

- *Addressees.* Standards can be designed as Business-to-Business (B2B) or Business-to-Consumer (B2C) concepts. The former are not communicated by means of a label to the consumer but only in intercompany transactions as quality signal (e.g. GlobalGAP). B2C standards address in contrast to this the end consumer by means of a label to evoke their willingness to pay more for special product or process characteristics (e.g. organic, fair trade, increased animal welfare standards). The rare combination of B2B and B2C orientation is united in the QS scheme.
- *Objectives.* Certification systems pursue two different aims: on the one hand, they are to make sure that minimum standards are adhered to. On the other hand, they serve to differentiate products in competition. For certification systems safeguarding minimum standards (e.g. QS scheme, IFS Food, GlobalGAP), quite often a strong orientation towards legal provisions, DIN standards, industry-specific guidelines and other (industry) standards is characteristic. However, differentiation measures are targeted towards creating an offer considered by the customers as special or even unique (e.g. organic production in case of organic standards; increased animal welfare standards)

(Porter, 1999). Thus, customer loyalty is to be increased resulting in the willingness of the purchasers to pay more so that the cost caused by differentiation are exceeded or at least recovered.

- *Focus.* Quality science differentiates between system, procedure or process and product audits (Pfeifer, 2001) (Chapter 5.3, Table. 5.4). System audits, partly complemented by individual procedural aspects, are quite often characteristic for those certification systems assuring upholding of minimum qualities in the market (QS scheme, globalg.a.p). The production processes are in contrast to this, however, in the centre of certification systems in the field of organic agriculture. The product focus is finally, amongst others, characteristic for DLG awards.
- *Geographical expansion.* Certification systems can be represented regionally (e.g. certification systems of the states), nationally (e.g. KAT – Verein für kontrollierte Tierhaltungsformen e.V.) or internationally (e.g. GlobalGAP, IFS Food, organic standards).
- *Number of participants.* The number of certified farms or firms ranges from a few dozen up to several 10,000 companies. In Germany, the QS scheme is the leading system and, in 2011, more than 130,000 companies from different stages of the value chain participated in this scheme.
- *Scope.* The scope describes the number of stages of the value chain covered by the standard. While some certification standards only address agriculture (e.g. globalg.a.p) or only processing (e.g. IFS Food Standard), other systems include several value stages (e.g. organic standards) or even all value stages (e.g. QS scheme).

The agri-food industry can no longer do without certification systems. It has become practically impossible for processors without certification according to the IFS Food standard to supply the German food retail. The QS scheme, in turn, covers in Germany between 95% of pork and 70% of beef production. Table 5.3 provides an overview – by no means exhaustive – of certification systems established in Germany. In addition, especially for exports, enterprises in the agri-food business are also certified according to foreign standards, for example the British BRC Global Standards for Food Safety.

The term sustainability is based on the three dimensions of organic, economic and social sustainability. Achieving a change of thought towards sustainable development in the society and companies is pushed by social and organic mega trends (BMU, 2010). Above all, sustainable value chains organised in a resource-efficient way are in the centre of interest (Trienekens *et al.*, 2009). Several authors stress that consumer acceptance of food products coming from animal origin are increasingly seen in conjunction with the aspects of animal welfare, environmental impact, health and food quality (Blokhuis *et al.*, 2008; Collins and Wall, 2004; Von Borell *et al.*, 2001; Wognum *et al.*, 2011).

Table 5.3. Selected certification systems in the German agri-food industry[1].

	Visible on the product	Not visible on the product
Standards to ensure process and product quality of foodstuff and feedstuff	QS; KAT (poultry)	BQM Landwirtschaft; IFS Food; BRC; GlobalGAP; GMP+; FAMI-QS
International standards on sustainability	MSC (for fish); ASC (for aquacultures); ProPlanet (REWE)	REDcert (for biomass)
National standards on sustainability		KSNL Kriteriensystem; Nachhaltige Landwirtschaft (Thür. Landesanstalt für Landwirtschaft); Zertifizierungssystem für nachhaltige Landwirtschaft (DLG); AgBalance (BASF); Sustainability Radar (BayerCropScience)
Organic farming associations with specific product and process guidelines	Biokreis; Bioland; Biopark; Demeter; Ecovin; Naturland; Neuland; Biosiegel according to EU regulation on organic production	
Standards for conventional products with specific characteristics	Neuland	
Label of origin	Bio-Zeichen Baden-Württemberg; Geprüfte Qualität Bayern; Geprüfte Qualität Hessen; Geprüfte Qualität Thüringen; Gesicherte Qualität Baden-Württemberg; Hergestellt und geprüft in Schleswig-Holstein; Vom Saar Landwirt; Unser Land	
Certification mark (determined by individual testing institutes)	Fresenius-Qualitätssiegel; Lebensmittel TÜV geprüft; DLG Prüfzeichen	

[1] DLG: Deutsche Landwirtschaftsgesellschaft; QS: Qualität und Sicherheit (Quality and Safety); IFS: International Featured Standard; BRC: British Retail Consortium; FAMI QS: European Feed Additives and Premixtures Quality System; MSC: Marine Stewardship Council; ASC: Aquaculture Stewardship Council; BQM: Basis Qualitäts Managementsystem (Basic Quality Management system); DLG: Deutsche Landwirtschafts-Gesellschaft e.V.; KAT: Verein für Kontrollierte Alternative Tierhaltungsformen e.V.

Here, a series of new standards is developing which have already been mentioned in Chapter 5.2. At the moment, the Global Reporting Initiative (GRI) index offers the most comprehensive approach to assess sustainability on the basis of economic, organic and social performance indicators (www.globalreporting.org). It is targeted, however, towards assessment requirements at corporate level. Suitable assessment approaches at the level of the value chain are still in the pilot phase.

A suitable concept has been developed, for example, by the REWE Group with scientific support, the so-called 'hot-spot analysis' (ProPlanet (REWE)). In connection with this concept, social and organic aspects from the raw material up to retail are evaluated according to goods groups and general hot spots are identified. In addition to scientific assessment, stakeholders from all social groups are included in the evaluation (Brinkmann and Petersen, 2011).

At the moment, companies are increasingly using standards and labels to offer rewards for activities and self-commitment in the area of sustainability. In the following, a small selection is shown.

ProPlanet developed by the REWE group is an important sustainability label in food retail (www.proplanet-label.com/de). During the awarding process, five defined steps have to be passed. For the labelling of their products, suppliers must carry out improvement measures for certain hot spots. Such trade-driven initiatives confront participants of the food producing chain with the major challenge to orient their own sustainability strategy towards customer requirements. They must organise and optimise organic and social sustainability criteria verifiably in order to participate actively in the market.

The Marine Stewardship Council (MSC) founded by the WWF (World Wide Fund for Nature) substantiates sustainable fishery. In the paper and packaging industry, there exist various approaches to certify sustainability, e.g. the FSC label (Forest Stewardship Council). Such a label is for example very popular with milk cartons.

In the area of certification of sustainably produced biomass, there exists the established approach of the ISCC certification system (International Sustainability & Carbon certification). Since 2010, the company REDcert is in possession of an approval of the German Federal Office for Agriculture and Food (BLE) to certify sustainably produced biomass according to the German biomass sustainability regulations (BioSt-NachV, 2009).

The Global Reporting Initiative (GRI) is a leading management concept for sustainability. The management concept is based on the Sustainability Reporting Framework. According to the G3 guidelines conceived, a company-specific sustainability report is prepared. The GRI index contains initiatives to reduce greenhouse gas emissions, such as the approach of carbon footprint. Amongst others, the MetroGroup and Westfleisch eG are using this concept of sustainability assessment. However, with the management concepts currently available on the market, the assessing component has been implemented more strongly than the controlling and regulating aspect.

Despite the wide-spread distribution of certification systems, they are still assessed with mixed feelings by the companies in the agri-food industry. Exemplary for this, the study

Box 5.3. GlobalGAP.

GlobalGAP was founded in 1997 still under the name of EUREPGAP by members of the Euro Retailer Produce Working Group (EUREP) on the occasion of increasing requirements of consumers on safety and sustainability of foodstuff and the lacking harmonisation of the standards developed by different bodies accordingly. GlobalGAP is a certification system for conventional agriculture including the areas of food safety (117 control points), environment (including organic diversity; 50 control points), traceability (46) and working conditions (21). GlobalGAP has thus agreed to ensure certain minimum standards in the market. Apart from food retail, GlobalGAP is nowadays also supported by other organisations from the areas of agriculture, processing and pre-processing industry. In 2011, altogether 112,600 agricultural enterprises in 112 countries were certified according to GlobalGAP. Out of them, 30% were certified individually; 70% had decided in favour of group certification. 74% of the certified enterprises are situated in Europe, 11% in Northern and Southern America, 9% in Asia, 4% in Africa and 2% in Oceania. Predominantly enterprises in the fields of fruit and vegetable production are certified, although further standards, such as for mixed feed production and handling of animal transports exist. GlobalGAP does not use any label; it is thus a rather typical B2B standard. The end consumer cannot recognise if a foodstuff is certified according to the GlobalGAP standard (www.globalgap.org).

of Schulze *et al.* (2008) regarding the attitudes of foodstuff processers to the IFS Food standard (at that time still International Food Standard) is to be highlighted. The authors of this study describe three different groups of companies:

1. Satisfied companies, who are aware of the high cost for certification but are convinced that external audits make sense and have a positive attitude towards the standard.
2. Unsatisfied companies, who complain about an unfavourable cost-benefit ratio of certification and have put up with audits only reluctantly.
3. Indifferent companies, who are basically satisfied with the standard but are not of the opinion that certification brings them any special advantages.

Similar three-fold divisions have been determined empirically in other studies on the IFS Food standard and also on other systems such as the QS standard as well (Gawron and Theuvsen, 2009; Jahn *et al.*, 2003).

Harmonisation effects and activities

Certification systems are based on the idea of verifying companies within the framework of neutral third-party audits in order to find out if they have adhered to certain requirements laid down in the certification standard (Meuwissen *et al.*, 2003). Therefore, certification systems have a harmonisation effect regarding the layout of quality management systems

Box 5.4. IFS Food.

IFS Food has been conceived in 2003 under the name of International Food Standard by member companies of the German Hauptverband des Deutschen Einzelhandels (HDE) and by the French Fédération des Entreprises du Commerce et de la Distribution (FCD) for the certification of own trade brands. Since the development of Version 5 of IFS Food in the years 2005 and 2006, the Italian trade associations were included as well. IFS Food is nowadays part of a standard family encompassing altogether six standards or guidelines. The development of IFS Food took place on the occasion of increasing consumer demands, rising risk of damage claims, globalisation of goods streams and the request for avoiding exhaustive multiple audits.
IFS Food serves as a quality and food safety standard for all processing stages downstream of agriculture. It encompasses in particular the following aspects: corporate responsibility, systems for quality and food safety management, resource management, manufacturing processes, measurements, analyses, improvements and food defence, i.e. protection of the food chain from acts of terrorism. The IFS Food standard is meanwhile used by about 11,000 food processing operators; until 2011, IFS audits have been carried out in 96 countries (www.ifs-certification.com).

Box 5.5. Bioland.

The beginnings of organic farming can be traced back to the early 20th century. In the 1970s, 'Bioland' was established as the name of the association and as a trademark. Today, more than 5,400 farmers, gardeners, winemakers and beekeepers farming together more than 659,819 acres are joined in Bioland e. V. More than 900 processors – mainly bakeries, butcheries, dairies, grain mills, fruit juice producers and breweries – process their products produced according to the Bioland guidelines. The guidelines contain, on the one hand, basic provisions for example for completely refraining from using genetically modified organisms, selection of the right location and protection of air, soil and water. On the other hand, the guidelines include rules for the areas of horticulture (crop rotation, fertilisation and humus management, plant protection, fertilisation, etc.) and animal production (rearing requirements, treatment of animals, transport and slaughtering, keeping livestock and imported feed, feeding, etc.) (www.bioland.de).

and of processes and procedures, for example documenting incoming goods or guaranteeing traceability of products.

Standardisation of processes is next to formalisation, i.e. a high degree of written form, a central characteristic – in the scientific sense (Weber, 1922) – of bureaucratic organisational forms (Adler and Borys, 1996; Mintzberg, 1979). Investigations by way of example of DIN EN ISO 9001 have shown that practically all features of bureaucratic organisational forms – goal orientation, a multitude of written rules, comprehensive written documentation, impersonality of rules, strong functional specialisation, specialised training of task managers – can be documented in certification standards (Theuvsen, 2004). Certification systems are therefore a prime example for bureaucratic tendencies in industry.

The bureaucratic nature of certification systems is by no means a coincidence. Academia has argued that bureaucratic systems indeed have negative aspects as well (Derlien, 1992), but these are countered by varied positive effects such as a high degree of dependability and reliability (the so-called 'technical superiority' of bureaucracy) (Weber, 1922). A central effect accompanying standardisation and formalisation is the support for learning processes of organisations. Standardised procedures thus serve as 'a store of knowledge' for best practices and therefore assure rapid implementation and broad application of improvements not only in companies but also in other areas (Adler and Borys, 1996). The bureaucratic elements of certification systems and the commitment of the system participants to achieve a best practice standard related with them, thus enables a continual improvement process – supported by applying the PDCA cycle (Hummel and Malorny, 2002). In this way, the quality capabilities of processes are increased at the same time (Pfeifer, 2001). A typical example for this positive effect of bureaucratic organisational forms is early implementation of *Salmonella* monitoring in the QS scheme, the lack of which has already for a long time been recognised as a weak point of German pork industry. With the QS scheme, for the first time ever a uniform procedure for executing *Salmonella* monitorings was established and thus implemented within a short period of time as best practice standard in major areas of the German pork industry. Due to the later enforcement of the pork *Salmonella* regulation, this effect has been intensified with the help of the legislator.

The harmonisation effect of certification systems is not only based on the adoption of a standard and verification if the system participants have complied with this standard. The often obligatory training activities result in approximation of procedures as well. The QS scheme for example provides for regular hygienic trainings in the area of meat processing. And finally, database-supported control of the certifiers by the standard providers also aims at guaranteeing quality and standardisation of procedures, in this case of audits (Schulze *et al.*, 2007).

Certification systems not only have a standardisation effect in the participating enterprises, but also create a need for harmonisation. The reason for this is the number of systems implemented resulting especially in internationally operating companies in multiple certifications on a regular basis and thus relevant cost burdens. In case of labels in the business-to-consumer area, in addition, possible confusion and excessive demands placed on the consumers owing to the multitude of information on the products which cannot always be evaluated confidently by them, is pointed out (Wissenschaftliche Beiräte für Verbraucher und Ernährungspolitik sowie Agrarpolitik des BMELV, 2011). To counter this problem, various ways can be followed:

- Formulation of public certification standards if a multitude of private initiatives resulted in fragmentation of a market and owing to increasing quality uncertainty possibly even to a (partial) market failure. One example of this is the creation of a uniform organic standard by the EC basic regulation on organic production No. 834/2007 (EU, 2007).
- International mutual benchmarking of standards if, first of all and independent of each other, in closely interrelated markets certification systems with similar objectives were established. One example of this are the coordination activities between the German QS, the Dutch IKB and the Belgian Certus system as well as the Danish Meat Association within the framework of the European Meat Alliance (EMA).
- Alternatively to harmonisation of standards, simultaneous verification of compliance with similar certification systems can be carried out within the scope of combination audits. So-called GAP audits are offered for agriculture for example enabling simultaneous certification according to QS and GlobalGAP standards.

A further contribution to harmonisation is made by the increasing internationalisation of certification systems. This is done, on the one hand, at the level of ownership of the standards. Correspondingly, the IFS Food standard was established in the first instance with orientation along the BRC Global Standards for Food Safety by companies of the German and French food retail. Later on, the Italian food retail followed the system as well and thus contributed to further internationalisation. Internationalisation is also achieved by establishing standards controlled from within one country but resulting in certification of system participants in various countries as is the case with GlobalGAP. This strong international orientation of a standard towards different countries of origin of agricultural products also results in harmonisation of quality assurance in those countries where certified enterprises are located. And finally, originally national standards also internationalise when certifying foreign suppliers intending to supply to the home market of the standard. This development can be observed in connection with almost all originally national certification systems for the agri-food industry. For example, in Central, Eastern and Southern Europe, meanwhile various producers and processers are certified according to BRC Global Standards, the IFS Food standard or the standard of the QS scheme (Gawron and Theuvsen, 2009).

5.3 Further development of sector-specific accreditation and certification systems

S. Lehnert[1], S. Bruckner[2] and L. Theuvsen[3]
[1]University of Bonn, Katzenburgweg 7-9, 53115 Bonn, Germany
[2]AGRIZERT, Zertifizierungs GmbH, Vorgebirgsstr. 80, 53119 Bonn, Germany
[3]Georg-August-University Göttingen, Department of Agricultural Economics and Rural Development, Platz der Göttinger Sieben 5, 37073 Göttingen, Germany
susanne.lehnert@t-online.de

Certification and accreditation are aiming at removing obstacles to trade in global markets effectively. Therefore, these objective confirmations are also referred to as conformity assessment. Trust in certificates, inspections, tests or calibrations stands and falls, however, with the competence of the party carrying out assessment. Many of these conformity assessment institutions, also called certification bodies, therefore prove quality of their own work by accreditation. Accreditation is a 'procedure to formally recognise the competence of an organisation to carry out specific tasks' [Definition from the DGQ publication 11-04 'Terms for quality management'] (DGQ-Schrift 11-04).

History

In contrast to various other countries, Germany did not have until the end of 2009 a single national accreditation body but a fragmented accreditation system consisting of about 20 private and public accreditation bodies in partly overlapping areas. On 1 January 2010, the Deutsche Akkreditierungsstelle GmbH (DAkkS) began with its work (www.dakks.de). With the establishing of this national accreditation body, Germany fulfils the European requirement according to regulation (EC) No. 765/2008 (EU, 2008) on standardisation of the accreditation system.

The German accreditation body DAkkS was founded in 2010 by the BMWi, the BDI and various German states. The DAkkS accredits testing, monitoring, calibration and certification bodies as well as laboratories. It is operating in the following areas:
- testing laboratories;
- calibration laboratories;
- medical laboratories;
- products;
- management systems;
- persons;
- inspection bodies;
- ring test providers;
- reference material manufacturers.

Box 5.6. Accreditation.

Confirmation by a national accreditation body that a conformity assessment body fulfils the requirements laid down in harmonised standards and, possibly, additional requirements including such in relevant sectorial accreditation systems to carry out a special conformity assessment activity (EU, 2008).

The DAkkS is working sovereignly, is subordinate to German administrative law and carries out accreditation tasks. This step opens a new chapter in the German accreditation system bundling the experience and competences gained over several years of the roughly 20 bodies in a new structure.

Within an accreditation procedure, certification bodies or laboratories have to prove to an independent accreditation body that they carry out their tasks competently adhering to legal and normative requirements and on an internationally comparable level. The accreditation body examines and monitors the management system and the competence of the assigned personnel of the certification body. There exist different standardisation foundations for the various products, personnel and management system certifications whose requirements have to be mapped in an independent management system of the certification body. Accreditations thus essentially contribute to guarantee comparability of conformity assessment results and generate trust in the quality and safety of products and services.

The national accreditation body must be completely independent of commercial pressures and is not allowed to work on a profit basis (Figure 5.8). It fulfils sovereign tasks and must therefore be authoritative and impartial. Competition with other accreditation bodies and accredited certification bodies must be excluded. It is not allowed to carry out conformity assessments in the area in which it accredits other bodies.

In order to guarantee a high level of conformity for the way of working of the accreditation bodies, the European legislation EU Regulation 765/2008 Article 2.16 (EU, 2008) determines that the accreditation bodies also shall subject themselves to an assessment – proven by peer evaluation.

At European level, the European Cooperation for Accreditation (EA) has been nominated by the EU Commission to carry out accreditation of testing, monitoring, calibration and certification bodies (Figure 5.9). On an international level, two organisations are operating: the ILAC (International Laboratory Accreditation Conference) has taken over

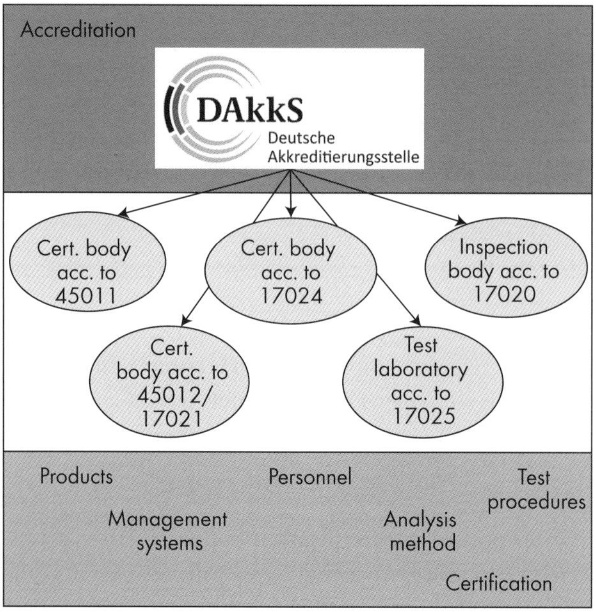

Figure 5.8. Accreditation foundations.

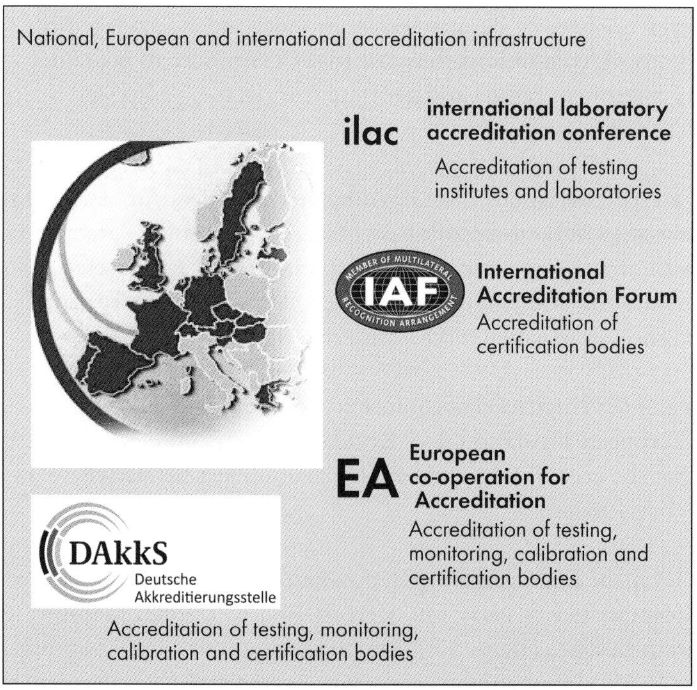

Figure 5.9. National, European and international accreditation infrastructure.

Box 5.7. Certification bodies to assess quality management systems.

- are approved for certain industrial fields (scopes);
- must be independent and impartial;
- must document their certification activity transparently;
- must have a certain legal form and corresponding organs;
- must provide auditors for the approved industrial sectors;
- must fulfil the requirements of DIN EN 17021.

international accreditation of inspection bodies and laboratories, and for accreditation of certification bodies at this level, the IAF (International Accreditation Forum) is responsible.

Certification – procedure and models

Certification is a measure taken by an impartial third party to show that adequate confidence exists in a properly labelled product, process or duly designated service being provided in accordance with a certain standard or another normative document (according to DIN EN 45020).

Companies operating in the agri-food industry have different possibilities to monitor quality-relevant business processes. Internal audits are arranged by the organisation itself to verify if the procedures and processes determined can be put into practice. This enables them to make statements regarding their efficiency and identify improvement potentials. They are also called first-party assessments. External audits take place with participation of external assessors or auditors; these can be, on the one hand, customer or supplier audits (second-party assessments) or certification audits, on the other hand. A certification audit belongs to third-party assessments since it is carried out by an independent party (Figure 5.10).

In case of external audits, a differentiation is made between system, process and product audits (Table 5.4). Within the framework of certifying management systems on the basis of different standards, so-called 'system audits' are carried out. When doing this, it is verified if the entire management system of a company including its procedure, work, process and specification descriptions corresponds with the underlying standard. Weak points and corrective and improvement measures are determined on the basis of:

- possibly laid down set of quality management rules;
- quality management manual;
- all process instructions;
- all specifications/instructions (e.g. recipes, raw material, finished product, storage specifications);

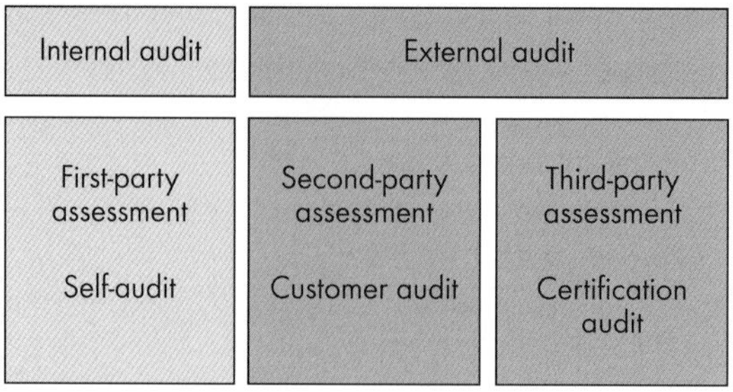

Figure 5.10. Different types of audits (Hansen and Almeida, 1993).

Table 5.4. Overview of different types of audits.

Type of audit	Purpose	Contents	Examples
System audit	System verification	Assessment of the efficiency of the entire system	DIN EN ISO 9001 DIN EN ISO 22000 DIN EN ISO 31000
Process audit	Process assessment	Assessment of the process capability and process efficiency Verification of requirements and proof documents	QS certification seal TIGA health standard DLG testing seal 'Sustainable agriculture' Eco labels
Product audit	Product verification	Control of quality requirements placed on the product	DLG quality label Product tests

DLG = Deutsche Landwirtschaftsgesellschaft; QS = Qualität und Sicherheit (Quality and Safety); TIGA = Tiergesundheitsagentur eG.

- all test documents;
- procedures to determine quality cost.

Accredited certification bodies are required to use auditors with a qualification according to DIN EN ISO 19011 in order to carry out this wide range of auditing activities with correspondingly qualified personnel (Kampa, 1996). These auditors have to prove specialised knowledge in quality management, profound knowledge of the industry and professional experience (Table 5.5).

Since the mid 1990s, industry-specific certification bodies such as AGRIZERT, Ifta Zert, LGA have been established with the objective to offer cost-effective and practically oriented conformity tests. It is characteristic for these sector specific certifiers that their auditors need to know, assess and prove the procedures and processes in the agri-food industry.

Table 5.5. Auditor qualification according to DIN EN ISO 19011.

Function	Prerequisites	Auditor training/ experience	Maintenance of qualification (in 3 years)
Auditor	Secondary education according to DIN EN ISO 19011 5 years of professional full-time experience with at least 2 years of experience in the field of QM or EM	Auditor education (at least 40 hours of training) 4 complete audits + at least 20 days of audit experience (of these 11 days on site) under supervision Assessment by certification body	Yearly participation in the exchanging of experience Participation in at least 2 external audits within 3 years Every 3 years, assessment by the certification body including monitoring
Audit team leader	Same as auditor	Same as auditor In addition at least 3 complete audits and at least 15 days of audit experience as audit team leader under supervision	Same as auditor
Specialised expert	Completed professional education At least 4 years of full-time professional experience in a specialised area	Qualification has to be proven before start of work Knowledge of procedures in certification body	Every 3 years assessment by certification body

Sector specific certifiers who are supported by specialised and professional associations and organisations have the advantage that their specialised auditor teams are able to harmonise the state of the art with the interpretation possibilities of the set of rules adapted to practice (Lehnert, 1995a).

5.4 Certification models in the agricultural sector

S. Lehnert[1], S. Bruckner[2] and L. Theuvsen[3]
[1]University of Bonn, Katzenburgweg 7-9, 53115 Bonn, Germany
[2]AGRIZERT, Zertifizierungs GmbH, Vorgebirgsstr. 80, 53119 Bonn, Germany
[3]Georg-August-University Göttingen, Department of Agricultural Economics and Rural Development, Platz der Göttinger Sieben 5, 37073 Göttingen, Germany
susanne.lehnert@t-online.de

It is typical for the food industry that mutual intensive integration of the preceding or the following stage in the company-specific quality management by agreeing on product and performance-specific quality objectives between agricultural primary production and food processing has to be achieved. The advantages brought about by this approach are quite obvious: the more precisely the following stage defines its requirements, the better the previous stage can plan, calculate and subsequently produce without any interruptions and economically optimised.

Within the value chains of the agri-food industry, an increasing integration pressure develops for agricultural enterprises. To meet these demands, the producing companies have to use modern leadership and management instruments (Lehnert, 1994a). In agricultural primary production, basically proof for process capabilities and procedural effectiveness by audits on site and supporting documents, such as QS certification seal or REDcert sustainability certificate, are required.

Suppliers of agricultural enterprises, such as manufacturers of feedstuff and production means as well as service providers, have mostly designed their own certified QM systems. This means for agricultural enterprises and associations being between these two stages in the value chain, that, on the one hand, they have to define their quality features for suppliers clearly, and on the other hand, have to know exactly about the customer requirements placed on them or prove their fulfilment.

Introduction and maintenance of QM systems because the customer demands these is nowadays playing an important role as underlying motivation for the enterprises of the German agricultural industry. Only if the agricultural producer stage is integrated in the value chains of milk, meat and vegetables, legal requirements – especially regarding origin assurance and consumer protection – can be fulfilled. The holistic and customer-oriented approach is aiming at involving all actors participating in the value chain in a multi-dimensional quality understanding (Lehnert, 1995b). Permanent and sustainable improvement of operational procedures is deemed to be an incentive for the establishment of QM systems especially for producer associations or enterprises with several production sites (Table 5.6).

Table 5.6. Reasons for the establishment of quality management (QM) systems (Lehnert, 1998).

Motive	Positive aspect	Negative aspect
Fulfilling customer requirements	Concrete customer requirements regarding the QM system Acceptance by customers Safeguarding the sales market	Reaction instead of action and own initiative Time pressure Investment pressure
Fulfilling legal requirements	Risk minimisation Analysis of legal conditions	Extensive documentation Security system
Assuring competitive advantages	Safeguarding the future Market orientation QM system also used as marketing instrument High level of personal initiative	Focus on certification Hardly any improvement effects No own objectives
Assuring operational procedures	Intensive analysis of business operations Increasing efficiency High level of self-motivation	Concentration on organisational areas with vulnerabilities

For producer associations, bundler organisations and network coordinators quite often the following aspects are decisive for establishing a QM system:

- improved control of production procedures;
- optimisation of customer-supplier relationships;
- uncovering of performance reserves for production and service;
- compiling a suitable documentation of procedures throughout the whole organization.

The following is expected as main external benefits:

- traceable process steps for partners in the value chain;
- efficient use of feedback from subsequent processing stages;
- traceable measures to guarantee product safety already during primary production.

Box 5.8. Bundler organisations – service providers for a safe acquisition chain.

In the QS inspection system for foodstuffs, bundler organisations are the first point of contact for agricultural and horticultural enterprises, animal transporters and food retailers. A bundler acts as the connecting link between the companies and QS. The bundler organises implementation of independent controls (audits) in the enterprises looked after by him and enforces participation in the obligatory monitoring programs.

For agricultural enterprises intending to introduce QM systems, four basically different initial situations are conceivable (Figure 5.11):

Variant 1: Individual enterprise with special branch of production, such as piglet production, asparagus cultivation, winegrowing.

Variant 2: Group of enterprises in an independent association with the above-mentioned branches of production.

Variant 3: Producer association with fixed structures (management board, executive board).

Variant 4: Producer association with quality program in vertical integration (e.g. marketing programs, quality programs).

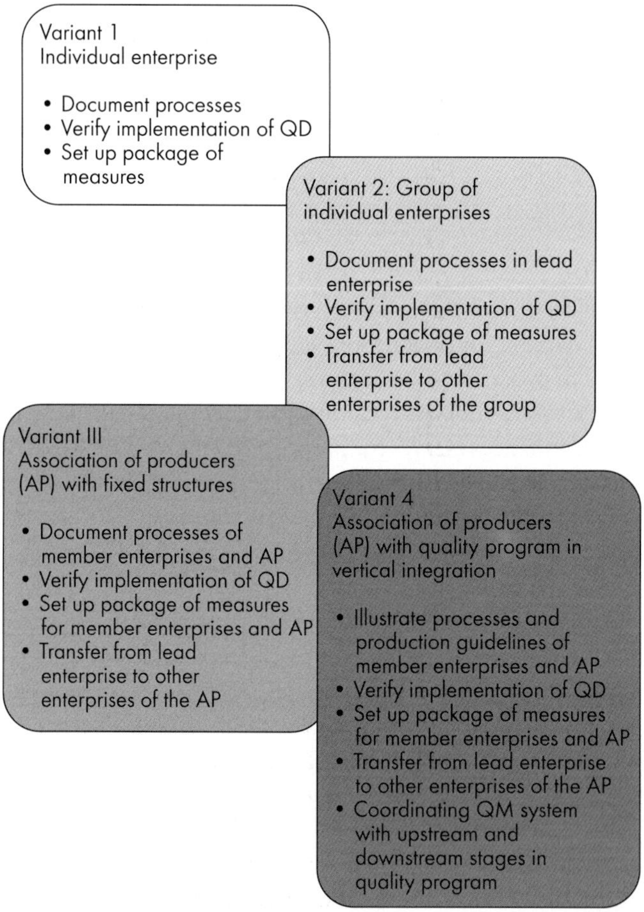

Figure 5.11. Four initial situations when establishing and certifying quality management systems (Lehnert, 1998).

If an individual enterprise decides to establish a QM system, the main phases are structured as follows: the enterprise-specific business or production areas of an individual company and all processes from purchase to sales are illustrated by the enterprise; the QM system is verified in order to determine if the requirements of the underlying quality standard have already been implemented partly or completely, and subsequently, a package of measures is prepared to close the still existing gaps in the QM system.

Quite often, however, several enterprises of a region or type of production think about establishing quality management systems for their enterprise at the same time. The impulse to do so often comes from the outside, i.e. from the customer or within the framework of central-regional marketing. Taking these initial conditions into consideration, it might make sense that first of all a so-called 'lead enterprise' within a group establishes a QM system with external support which is taken over gradually by the other group member enterprises. In this regard, it is important to win managers for this plan who are very motivated and willing to pass on their work results and experience to colleagues (Petersen *et al.*, 1997).

A third initial variant assumes that the impulse to establish a QM system and the wish to have this system verified by a neutral party does not come from the agricultural individual enterprises but from bundler organisations. These are horizontal associations of the agricultural enterprises in the form of producer associations. They are managed with differing forms of associations:

- farming associations with own farming rules;
- supplier associations with or without obligation to supply;
- self-marketing organisations with their own businesses.

In bundler organisations, the individual producer association does not appear individually on the market. Instead, the bundler with his individual organisational structure places the offer of the association on the market. The member enterprises are not only suppliers within the producer association but also influence the determined production and quality rules. In bundler organisations, the different functions or elements of the quality cycle are distributed to bundlers and individual enterprises. Usually, marketing, sales, contract verification towards the customer and definition of a quality policy are the classic tasks of the parent organisation. The individual enterprises, above all, contribute process-related quality elements to the QM system.

A fourth initial situation results when all enterprises or companies of a collaborative, vertical value chain coordinate their stage-related QM systems and develop them chain-specifically. Together with trade organisations, production and processing represent a quality program and have a contractually determined customer-supplier relationship.

These different implementation models require agro-specific models for their certification. This is due to the fact that transfer of typical models with hierarchically structured levels of responsibility and a 'physical' organisation as represented by the classic corporate model only limitedly satisfies the needs of the agricultural industry. Although agricultural individual enterprises are quite frequently small and medium-sized companies, the typical system model 'individual enterprise' may apply nevertheless (Figure 5.12). In case of bundler organisations, it is necessary to offer modified solutions. Enterprises forming a group in order to establish a QM system mostly prefer a more cost-effective group certification. Another alternative for entire value chains is the model of quality associations.

With matrix certification, the quality management systems of economically and legally independent companies with several subsidiaries at different locations, e.g. filial operations or the systems of horizontal associations of economically and legally independent companies are certified as a legally independent organisation.

The individual subsidiaries/companies can be independent or dependent under company law, must, however, be subordinate to a common QM system. All locations must enter a legal or binding agreement with the identifiable central of the organisation. The QMS is laid down, introduced and monitored on a regular basis by the central body. This means that the central organisation is allowed to impose corrective measures if these are needed at one of the locations. The products or services offered by all companies/locations must be essentially of the same type and correspondingly be manufactured according to the

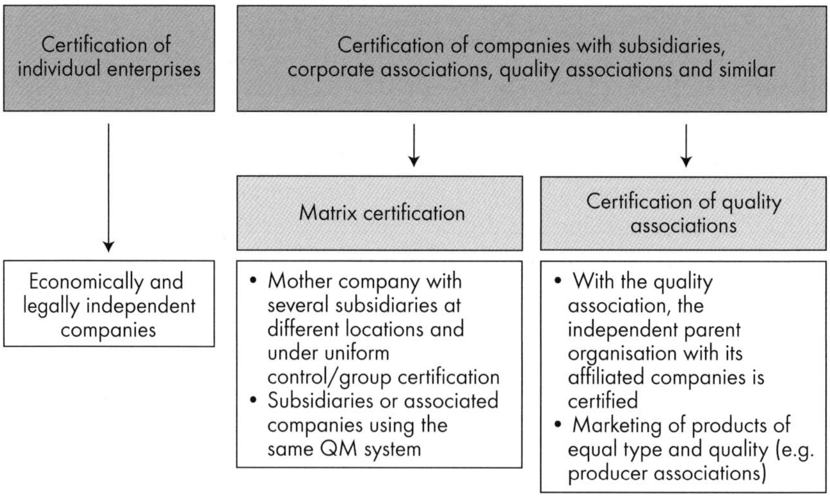

Figure 5.12. Certification models in the agricultural sector.

same methods and procedures. All relevant locations/companies (including the central administrative function) must be subject to the internal audit program of the organisation.

With the model of quality associations, quality associations (producer associations according to the German Market Structure Act) and horizontal associations consisting of economically and legally independent companies to form a legally independent organisation with the objective of marketing of products of the same type and quality, are certified.

The individual companies can be independent or dependent by company law, are subject, however, to a common QM system. All locations must enter a legal or binding agreement with the identifiable central of the organisation. The QMS is laid down, introduced and monitored on a regular basis by the central body. This means that the central organisation is allowed to impose corrective measures if these are needed at one of the locations. The products or services offered by all companies/locations must be essentially of the same type and correspondingly be manufactured according to the same methods and procedures. All relevant companies (including the central administrative function) must be subject to the internal audit program of the organisation. Referred to marketing and logistics and defining production and quality guidelines, the companies subordinate themselves to a uniform management.

There exists a central QM system at the level of the parent organisation. Legal enforceability of the QM system and further agreements within the organisation has been laid down by means of contracts.

The organisation of the quality association supports its independent members in their formation and management of the company with know-how and an extensive service offer. Cooperation is not limited in time.

For agricultural individual enterprises having several hired workers and/or several places of production, it will be in future initially a matter of extending existing QM systems for partial branch offices to all areas of production. In this way, they can explain the comprehensive quality capabilities of their enterprise to their market partners coming increasingly from the international economic area. System extension by environmental aspects, sustainability requirements and social responsibility is an important task in this regard. During a second development step, it will be important to bring concepts for integrated management systems including the field of activity of the intercompany chain management into much sharper focus in addition to customer, employee and process orientation within the corporate operational and organisational structure (Lehnert, 2000).

In the agri-food industry, in the past few years, a series of standards and labels have been established, the requirements of which have been realised in different combinations within organisations (Figure 5.13). Partly, these requirements of standards exist next to each other, and, to a certain extent, first steps towards their integration have been undertaken. Certification of combined management systems can be carried out within the framework of combined audits. When doing this, at a common audit date, verification of existing conformity requirements is coordinated and executed to keep the efforts for the organisation to be certified as low as possible and to minimise interruptions of operational procedures. Duration of the combined audit depends on the extent and number of the underlying requirements and standards.

From an organisational point of view, the following aspects support combined auditing:

- reduced administrative effort and cost;
- less interruptions for executive personnel;
- continuity of work and operational processes;
- less local disturbances.

For certification bodies, execution of combined audits is becoming more and more difficult, however, since the auditors used for them have to furnish proof of an accreditation for the special combination of standards. Since especially the standard owners in this industry

Frequent combinations of standards for certification	DIN EN ISO 9001	DIN EN ISO 22000	IFS	BRC	QS	GMP	REDcert	MSC	GlobalGAP	Organic (EU Eco reg.)
DIN EN ISO 9001			■	■	■	■	■			
DIN EN ISO 22000			■	■						
IFS	■	■			■			■		■
BRC	■	■								
QS	■		■			■	■		■	
GMP	■				■		■			
REDcert	■				■	■				
MSC			■							■
GlobalGAP					■					
Organic (EU Eco reg.)			■					■		

Figure 5.13. Combinations of certification standards.

usually place additional requirements on the auditors exceeding the basic requirements placed on the auditor qualification mentioned in Table 5.5, the qualification effort for auditors increases. Hence, for example special training measures have to be proven and/or tests regarding standard contents and possibilities for implementation have to be undergone. This leads to auditors specialising in certain standards and labels, and the certification bodies thus have an increased organisational burden in connection with combined audits which also leads to higher personnel cost for them.

Chapter 6. Food chain management systems

This chapter elaborates on food chain management issues and food chain management systems. Section 6.1 provides information and communication systems in food chains. In Section 6.2, traceability in food chains will be discussed. Both sections start with definitions and background information, followed by several case examples. Section 6.3 addresses innovation management in agro-food networks and discusses success factors for food chain innovation. Section 6.4 deals with the topical issue of cold chain management with a number of illustrative examples. Section 6.5 concludes by discussing environmental management systems in food chains.

This chapter will inform you about:

- information and communication systems in food chains;
- traceability in food chains;
- innovation management in agro-food networks;
- cold chain management;
- environmental management systems in food chains.

Keywords: information systems, transparency, open innovation, cold chain, innovation management

Summary

Food chain management encompasses the flow of food products from primary producer until the final customer. Effective information exchange is a pre-requisite for collaborative actions in food chains. Chain-wide information systems support transparency of products and processes for product identification (traceability) as well as for logistics and quality management purposes. Increasing demands of food chain stakeholders push chain actors towards joint innovations. In this regard, a topical area in food chain management is management of the cold chain in combination with innovative packaging strategies. A relative new field in food chain management is collaborative environmental management, e.g. focusing on optimization of waste streams.

DOI 10.3920/978-90-8686-789-9_6, © Wageningen Academic Publishers 2014

6.1 Information and communication systems in food chains

J. Trienekens[1], A. Czekala[2], V. Schütz[2], N. Wognum[1] and J. Denolf[1]
[1]*Wageningen University, Management Studies Group, Hollandseweg 1, 6706 KN Wageningen, the Netherlands*
[2]*DRV, Deutscher Raiffeisenverband e.V., Pariser Platz 3, 10117 Berlin, Germany*
jacques.trienekens@wur.nl

Supply chain information systems

To create value for consumers as well as for each actor involved, supply chains need to efficiently satisfy the changing demands of their customers. Enhancing efficiency and effectiveness requires the exchange of information throughout the supply chain. Actors need to be aware of the value of their product for any downstream actor including the end consumer. Information systems within and across supply chain stages are necessary to achieve a seamless information exchange.

Information exchange in the supply chain is essential to guarantee food quality and provenance to all users of food and food products. Intensified information exchange and integrated information systems involving all chain actors are needed for various reasons:

- Consumers require more and more information about the origin of products and the way a product is produced.
- Availability of operational quality and safety data across the chain supports adequate risk management.
- A reliable exchange of quality and health data throughout the chain provides actors with an instrument to plan their production and sales processes better and to match the right quality to the right market better.
- The increasing complexity of logistics flows caused by product differentiation, market segmentation and internationalisation, demands insight into production and stock data throughout the food chain, so that companies can make better forecasts and plan logistics and distribution processes more effectively.
- Based on the experience gained during recent crises in the global food sector, traceability has proved to be a key capability for companies to find the origin of problems and recall hazardous products quickly.

To meet these challenges, vast amounts of data along the food chain should be captured, stored and analysed to support the taking of decisions. In the food sector, barcoding, RFID (Radio Frequency Identification) and in future also biometrics (e.g. DNA technology) enable capturing of data in the operational processes of food supply chains. Pork and veal supply chains, for example, use RFID tags in ear labels to identify animals throughout the supply chain up to the processing stage. In recent years, the application of RFID in animal supply chains was extended to automatic identification (through readers) at feeding stations, enabling adaptation of feeding and medicine use (mixed with the feed) to the individual animal, automatic weighing and selection of animals, temperature measurement to identify sick animals, etc. RFID is also increasingly used in vegetables and fruit production and in processing companies to identify many products. This enables an adequate management of production and distribution processes and traceability and recall management. The use of DNA technology will be increasingly viewed as a promising solution to check provenance of food products.

Just like in other industries, data captured in operational processes are used for the planning and evaluation of production. In this respect, we can refer to examples such as Warehouse Management Systems (WMS), specifically designed for the food industry for quality-controlled storage and distribution processes; Laboratory Information Management Systems (LIMS) for quality and safety control of certain product lots or (groups of) animals; and the more generic Enterprise Resource Planning (ERP) systems for integrated planning purposes. Furthermore, systems to communicate traceability data are needed, for example, Electronic Data Interchange (EDI) or eXtended Markup Language (XML). Although most of the larger food companies in Western countries have access to these kinds of systems, real integrated information systems on supply chain level are not yet common. GS1 (www.GS1.com), the international association that develops and implements standards for use in logistics systems (unique coding of product entities, transportation entities, etc.), is a common and standardised infrastructure to facilitate efficient and effective exchange of data in the supply chain.

Electronic commerce facilitates commercial supply chain transactions, and this creates new opportunities for inter-company collaboration and access to new markets. Electronic internet-based communication systems can support the following, i.e. (1) identifying potential trading partners; (2) selecting a specific partner; and (3) executing the transaction. In the food sector, electronic auctioning is an example of electronic commerce. Product information (quantities, quality grades, delivery schemes) is entered into an electronic auction place where buyers can bid without being physically near the products. Also products are not transported to an auction building but can be shipped directly to the buyer. Benefits of buying or selling products through electronic auctioning include reduction of transaction costs and easy access to a large number of buyers and suppliers. However, food companies increasingly maintain direct trade relationships with their suppliers, thereby surpassing the auctions. In many cases, electronic auctioning nowadays is used to smooth the peaks and valleys of supply and demand.

Another development in the food sector is the emergence of joint chain information systems. In collaborative supply chain environments where there is close collaboration between horizontally and vertically linked companies in the food chain, a joint infrastructure consisting of networks, coding standards and integrated or coupled databases is established. Large companies design their own systems to communicate with suppliers and for analysis purposes. However, small producers increasingly start collaborating through an integrated information exchange supported by third-party service (software) providers. Integrated or coordinated databases at various network partners serve as the backbone of these integrated supply chain systems. The goal of these systems is the exchange of (quality performance and safety) data to enable transparency in the chain and to optimise processes throughout the chain. Chain-wide information systems require that actors are willing to share and communicate information. Safety, security, reliability, and ownership are important concerns that have to be addressed.

Moreover, the necessary process changes accompanying a chain-wide information system need to be understood. Such process changes have an impact on people's tasks, their responsibilities and authorities. If they are not well managed, this may result in resistance towards and even refusal of the information system. The organisation of a supply chain affects the complexity of implementation of chain-wide information systems. An integrated supply chain with a clear supply chain director, which may be a person or an association, has more power to more or less force actors to participate. Conversely, it is more difficult in a supply chain dominated by market relationships to motivate actors to participate and commit themselves to using the system. Definition of a shared goal as well as recognition of an individual added value are necessary to involve actors.

Although the information exchange between companies is increasingly facilitated and supported by sophisticated systems, it is difficult for consumers to gain access to encompassing information concerning product streams (e.g. origin of raw materials, transport routes), quality characteristics (e.g. biomarkers) or the type of processing used. The relevant signals conveyed to consumers are largely in the form of on-product labels and logos. These labels are usually representative of information clusters regarding controls, process organisations, product compositions, origin, etc. Intelligent ICT devices owed by consumers such as smart-phones may be linked with the respective label owners through e.g. the internet cloud for unlocking hidden information. In this respect, hidden information would not have to be restricted to label information but may include information from legislation as well as information on brands or brand owners.

In the next sub-sections, typical examples of supply chain information systems from the food sector will be presented. The first example is a health management system called Wikiporc, the second example shows how information is exchanged in the Dutch pork chain, and the third example describes the pig health management system TiGA (Tiergesundheitsagentur eG).

Wikiporc

Wikiporc (www.wikiporc.fr) is a web portal used by the pork sector of Group Glon and its partners. It is part of a technical and organisational system to manage animal health, quality and performance on a pig farm. This portal improves synergies between different operators in the pork chain regarding preventive and curative animal health and it integrates all data relevant for analysing a particular health situation on a farm and makes these data available for different users. The system assures confidentiality to its users.

Wikiporc currently represents 1,800 pig farms in Western France. It is web-based and provides information on pig and carcass results to farmers, farm technicians and veterinarians. These results provide them with a global and personalised view enabling them to improve their pig management processes and to achieve sustainable quality and performance.

Wikiporc was started as a system for veterinarians who were also involved in designing and developing the system. Wikiporc has been further developed as a supply chain information system for its leading French organisation in animal nutrition called Sanders (www.sanders.fr) and its partner, i.e. the farmers' cooperative Porc Armor (www.porc-armor.com). Many new users have been added to the system while interfaces with existing systems have been improved. Step-by-step functionalities have been enhanced. Advisors and veterinarians are provided with a more detailed overview of the health status of farms (health, quality, technical and economic issues). Pig farmers have user-specific dashboards with relevant information which is also available for their technicians and veterinarians.

Wikiporc has been implemented using the Chainpoint® technology which facilitates collaboration between business partners. The Chainpoint® database unites data from different sources while retaining the relationships between data. Chainpoint® is based on secure web technology, making local software application unnecessary. Wikiporc had a large impact on the pork sector in Western France because information exchange has been largely improved leading to better animal health management. The system can be viewed as an innovation with large potential for enhancing the way of working in the pork supply chain. When consistently and widely used, a system like Wikiporc can contribute to a so-called risk-based meat inspection programme. Such a programme reduces the number of audits based on performance results. So far, the Wikiporc system has been applied in the upstream part of the pork supply chain. Extension to downstream stages may be feasible in the long term. The Wikiporc system could then contribute to a programme to distinguish the Group Glon products in the eyes of the consumer. Specific branding may be an option.

Exchange of information in the Dutch pork chain

The Dutch government has defined many rules and regulations that also require the use and exchange of information. For example, farmers need to include information in a document that accompanies pigs ready for slaughter. This information is the so-called food chain information (VKI). Farmers need to record, among others, the number of animals, the health status of animals and the farm they come from, animal medicine use, analysis results important for food safety, and the name and address of the veterinarian involved. VKI is part of the quality management system IKB (Integrated Chain Control). Electronic communication of the information increases information reliability. The slaughterhouse receives the electronic information before the animals arrive. It needs to specify a HACCP protocol for dealing with the information, including communication with the food authority VWA. Without VKI information, animals cannot be slaughtered.

The information exchange across the supply chain enables benchmark and trend analyses regarding performance of animals, products and processes. These analyses support actors to improve their processes as well as the supply chain to advance value creation. An example of a chain information system in the pork sector is FarmingNet launched by Vion in 2005. It is a web-based information system that provides farmers with on-line access to data about the pigs they have supplied. Analysis of the data is performed by Vion which shows the farmers the quality level and degree of uniformity of their pigs, both of which influence their net profit. The system not only focuses on better planning and control of operational processes in the pork chain, but also on mid and long-term optimisation of various production and distribution processes. Its future use could be extended to achieve two potential additional advantages for the slaughterhouse-farmer link:

1. An overview of body and carcass deviations per batch/stable may provide a better insight into the influence of climate control on growth of the animal which again may lead to additional returns (increased pig growth and reduced throughput and cycle times), reduced costs (decreased deviations), and increased resource usage (energy).
2. Better use of information on animals to be delivered may also lead to improved accuracy of weight partitioning of pigs at the time of delivery related to pig pay-off. This information can result in reduced costs (increased optimisation of delivery weight) and increased harmonisation of market quality concepts.

The TiGA-standard – health management in pig production

The co-operative Tiergesundheitsagentur eG (TiGA; animal health agency) was founded on 12 October 2012 by five German pig producer and trader cooperatives. At the moment, the TiGA has eight member organisations. Two additional organisations use the TiGA-standard and its database. TiGA established the first nation-wide health standard for standardised monitoring and communication of the pig health status. The standard is based on regular farm audits, blood and excrement analyses and the presentation of the results in the TiGA database. The TiGA management board implemented a consulting committee consisting of experts from veterinary organisations, science, trader organisations and animal health services. These members focus on the process of harmonisation and further development of the standard.

TiGA works as a service organisation for the pig producers, the veterinarians as well as the producer organisation and livestock trading organisations. The aim of the TiGA activities is to achieve a sustainable high animal health status on farm, regional and production chain level. The TiGA system supports decision makers in the regulation of health management processes. Decisive for the regulation is the normative-actual value comparison and definition of the obligatory target values to keep the process in balance. Intervention in the health management processes are based on human decisions and activities. Animal health services in general can be organised according to Figure 6.1.

TiGA is involved in target setting through the use of different pathogens as health indicators (performance target in a broader sense) and the requirements of the certificate laid down in the TiGA standard (customer requirements). TiGA works as Netchain coordinator for the health management activities of the bundlers (combining farmers' activities) and the veterinarians according to the TiGA standard (TiGA, 2012). At the moment, the pig producing and trading organisations function as bundlers for their members and customers – the farmers. In general, the direct TiGA membership is open to other organisations and institutions as well as to farmers. The information and communication flow is organised according to Figure 6.2 and involves the TiGA management board, the bundlers, the veterinarians and the database.

Information and communication are important elements for the improvement of pig herd health and during crisis prevention and crisis communication. The system developed by TiGA works with the following elements:

- *Prevention*. Prevention is provided because the piglet producer gets regular (quarterly or half-yearly) and standardised information about the health status of his pig herd. In collaboration with the service providers – e.g. the bundlers and veterinarians – he can improve the health status on the basis of the results achieved. His customers – the fattening farmers – receive standardised information in the form of a certificate concerning the health status of the piglets. Accordingly, health management at the fatting farm or the assembling of piglet batches can be adapted to this.
- *Monitoring*. The guidelines prescribe compulsory and elective monitoring on specific pathogens. In addition, the database offers the possibility to include the monitoring of other individually elected pathogens. The database allows an analysis regarding the spread of a pathogen within a certain period.

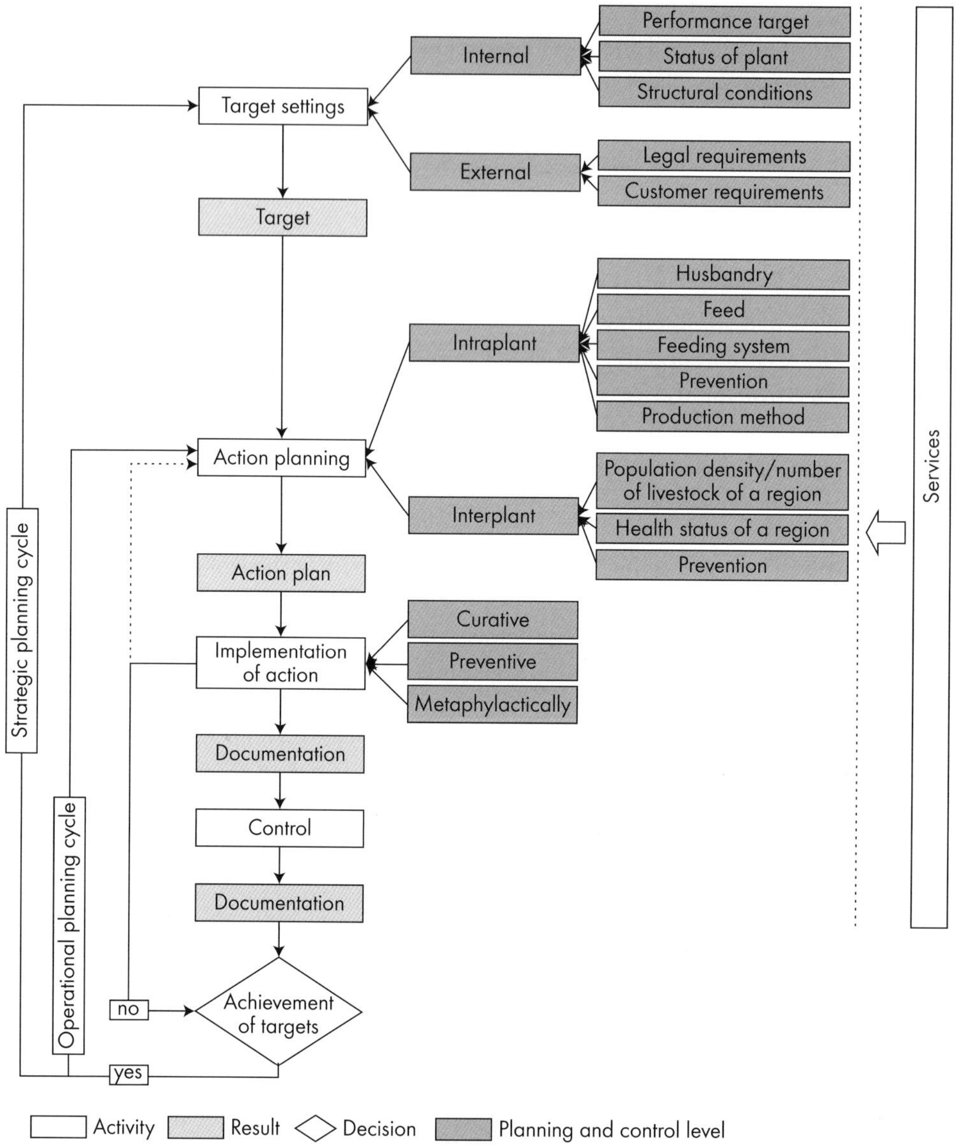

Figure 6.1. Flow chart for planning and implementation of fields and constitutive criteria of services for animal health management (Schütz, 2009).

- *Data exchange*. The infrastructure of the database enables regular data exchange between existing databases (e.g. laboratory systems) and the user of the TiGA system as well as data analysis at regular intervals. It also allows data transfer along the chain and a connection to other systems.

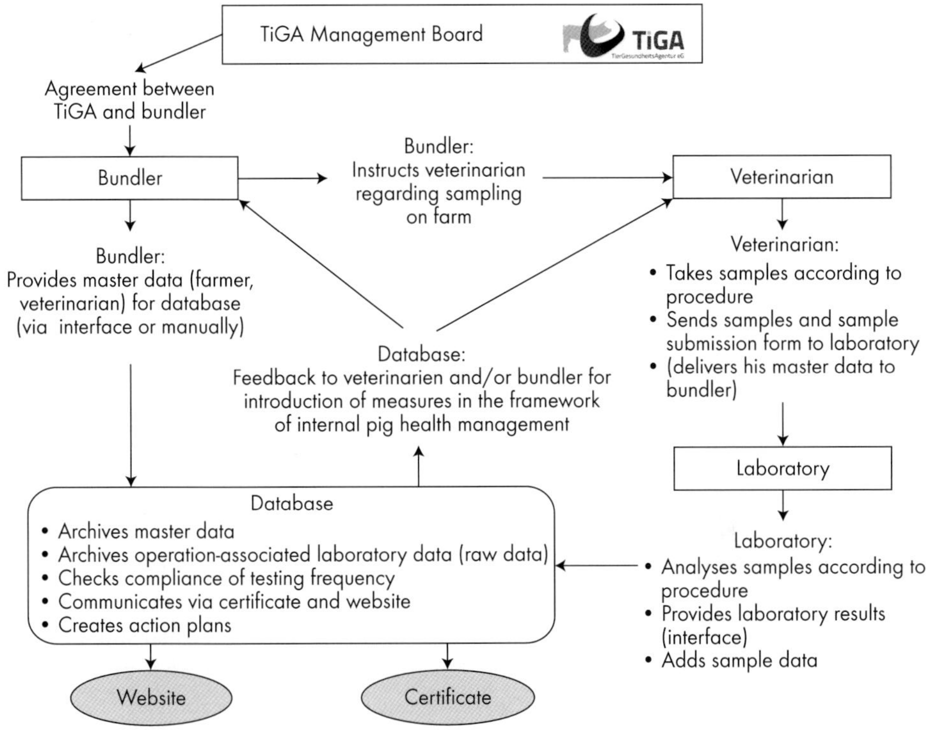

Figure 6.2. Operational structuring of participation in the Tiergesundheitsagentur eG (TiGA) Standard (TiGA, 2012).

- *Communication*. The system improves communication of the participants along the meat supply chain in general – not only on the basis of the database. Requirements laid down by the participants involved can thus be forwarded to the suppliers more easily. This is also useful in times of crises as for example epidemic or zoonotic diseases. A communication structure is already established. Communication is also possible with other stakeholders, e.g. in the sense of public-private partnerships.

The TiGA health management system for pigs is based on a standard. But the four elements (prevention, monitoring, data exchange and communication) are flexible regarding configuration of the relevant elements. Therefore individual requirements can be considered. This foundation, especially the flexibility of the database, enables variation and individual application-dependent use.

6.2 Traceability and trust

J. Trienekens[1], A. Czekala[2], V. Schütz[2], N. Wognum[1], J. Denolf[1] and L. Theuvsen[3]
[1]Wageningen University, Management Studies Group, Hollandseweg 1, 6706 KN Wageningen, the Netherlands
[2]DRV, Deutscher Raiffeisenverband e.V., Pariser Platz 3, 10117 Berlin, Germany
[3]Georg-August-University Göttingen, Department of Agricultural Economics and Rural Development, Platz der Göttinger Sieben 5, 37073 Göttingen, Germany
jacques.trienekens@wur.nl

Transparency of a chain is the extent to which all the chain's stakeholders have a shared understanding of and access to product-related information that they request without loss, noise, delay or distortion (Hofstede *et al.,* 2004). This definition implies that data must be relevant, accurate, factual, reliable, timely and available in an appropriate quantity. Moreover, quality information must be readable while the information exchange must be reasonable and properly arranged (Hofstede *et al.,* 2004). Apart from well-designed information systems, trust between partners is key to achieving transparent supply chains (Lindgreen *et al.,* 2005) leading to higher levels of loyalty and better formal and informal communication. One key characteristic of transparent food chains is that products are traceable from fork to farm. This section goes into the subject of traceability.

For consumers traceability of products is important to guarantee the origin of the product and the way it is produced. For the government, the main incentives to support food supply chain traceability include protection of public health through timely withdrawal of unsafe products, to prevent fraud regarding food product authenticity and to control livestock diseases (such as swine fever). The industry's motivation regarding traceability is four-fold: first, companies need to comply with differentiating demands from consumers as well as legislative demands. Second, when incidents occur, companies are required and want to be able to quickly recall products from markets or links downstream of the supply chain to limit the number of incidents and minimise costs. Third, by improving the information exchange through integrated information systems, optimisation of business processes will be much easier as product and process attributes can be coupled with process performance. Fourth, an important way for food companies to add value is by paying attention to and labelling products according to distinguishing intrinsic and extrinsic food product attributes. By implementing transparent and integrated information systems, organisations can improve their image as they can propagate that their products satisfy specific quality (intrinsic and extrinsic) demands. As already referred to earlier, other motivations are logistical optimisation, quality assurance and process optimisation.

Traceability systems imply identification and tracing functionalities. Product identification and product tracing each refer to a different set of requirements imposed on products and materials. The importance of identification is to be able to clearly distinguish one material or product from another one during the manufacturing process by means of tags, labels, routing sheets, colours, etc. Identification is typically done according to the supplier's established procedures. The basic idea of tracing is the possibility to determine where a certain item is located and to trace the history of that item. On the basis of that information, it should also be possible to determine the source of any (quality) problem of an item, and it should be possible to find out where other items with the same problem are located in the supply chain. Traceability is of importance at chain level as well as at company level. At company level, a system should provide information on the location of the product and on the history of the product (product and process information). At chain level, in addition to information on the location of products, information on the origin of the product is also of importance. In this regard, it is important to identify the current unique characteristics of lots (components) and the historical relationship between lots.

To enable transparent information exchange, detailed registration of process, resource and product characteristics, such as history of products, quality variation, etc., is essential for the food industry for the purpose of product differentiation, traceability and recall management, production management and for complying with new rules and regulations.

It is imperative that data collection and recording in the system take place as soon as possible, at least in time to allow for the necessary data quality and usage. A large actuality gap between the occurrence of events and recording of data is often a problem when systems have to be used more or less in real time as in the case of product recalls. In many cases it has been a (sometimes too) big challenge to timely and adequately link a food incident to data recorded earlier in time.

Product-related information in the context of food chains can refer to various types of information from different sources. All these types of information pertain to the smallest homogeneous product unit (Beulens *et al.*, 2005).

- *Inherent properties of a product*. These can be seen or otherwise measured in the product (i.e. intrinsic product attributes). Most inherent properties can change over time, e.g. taste, content of chemical components, bacteriological status, and visual attractiveness. Some are not likely to change, e.g. size.
- *Process properties*. These constitute the history of what has happened to a unit. If units resulting from different production processes with different resources are combined, for instance, only on the basis of equal inherent properties, units with non-homogeneous process properties are formed. This may become a problem, when tracing product provenance through the supply chain regarding process properties becomes necessary, e.g. in the case of a recall.
- *Properties of means of production used for the product*. This is in fact an extension of the previous category. It includes machines and labour. In the case of machines,

contamination issues can arise. For instance, if a machine is used for non-genetically modified (GM) seed after having been used for GM seed, very rigorous cleaning is necessary. In the case of labour, ethical issues can arise. Customers may, for instance, refuse to buy products if they suspect that child labour is involved.

- *Origin or provenance of product data*. This deals with information about the processes, associated actors, resources, raw materials and intermediate products that have been used to produce the product.
- *Actors involved*. They execute processes, 'produce' products, services and information. Properties of actors may be 'inherited' by the processes and products produced.
- *Relationships between the entities* previously mentioned with associated properties of interest. For example, if a process consumes inputs, uses machines and produces according to a certain specification, the properties may be associated with that combination of entities.

Transparency demands imply a number of requirements for information systems in food industries. We present two examples of data registration and processing requirements for companies in different food supply chains. The first example illustrates data requirements linked to processes in an average distribution company in the fruit and vegetables supply chain consisting of relatively simple storage, sorting and packaging processes (Trienekens and Van der Vorst, 2005) (Table 6.1).

Table 6.1 shows the importance of separating suppliers and supplier characteristics from different production lots. Quality differentiation and registration of quality differences during storage and processing/packaging is another information system challenge for such a company.

Table 6.1. Important data storage and processing requirements for a trader of fruit and vegetables.

Supply	• Storing of purchase specifications such as origin of product and supplier data (including production specifics) • Coupling of lot numbers with supplier data and production data • Coupling of lot numbers with entry control data and (differentiated) quality attributes
Storage	• Sorting and grading into new lots with specified quality attributes • Coupling of new lot numbers with storage location and conditions (e.g. temperature, time)
Assembly	• Registration of new lot numbers linked to preceding lot numbers. Production resources used
Packaging/labelling	• Coupling of lot numbers with label
Sales	• Coupling of lot numbers with invoice data • Coupling of lot number with distribution data

The second example depicts more complex information requirements for food industries (Table 6.2). Table 6.2 shows the complexity of separating suppliers and lots in companies with larger numbers of suppliers and more complex production processes.

Traceability and trust: a tricky relationship

Traceability is a means to improve the transparency of food supply chains. It is widely undisputed that, in this regard, trust plays a pivotal role (Lindgreen *et al.*, 2005). Nonetheless, a closer look reveals that the relationship between trust and traceability (i.e. a specific type of transparency) can be a tricky and even contradictory one (Frentrup and Theuvsen, 2006).

Table 6.2. Important data storage and processing requirements for various business processes in a food industry.

Production planning	Registration of product lot attributes
Order management (purchasing)	• Registration of data during order entry (supplier attributes, product origin, delivery data, delivery time, product attributes such as quality class, extrinsic product attributes or production according to standard, etc.) • Having insight into the location of the ordered goods in the supply chain • Registration in case of purchasing raw materials for the production of samples in R&D
Warehouse management	• Links between batch numbers of suppliers and lot numbers of food industry • Registration of data on lot or product characteristics during receipt • Registration of product or lot attribute data after splitting or mixing of lots as a result of sorting and grading or otherwise • Location control per lot and quality monitoring • Reverse logistics: identification of raw materials lots that are returned from production
Manufacturing	• Registration of actual lot numbers that are used in production • Registration of quality attributes per "batch" or individual item (e.g. an animal) • Registration of process variables per 'batch' or individual item • Lot traceability in case of using more batches for one packaging order
Order management (sales)	• Registration of actual lot numbers that are sent to the customers • Registration of complaints • Using shelf-life restrictions in sales
Freight management	• Registration of data during order picking and truck loading (employee who picked the order, actual lot numbers per sales order line, departure time, departure temperature, etc.); thereafter, monitoring and controlling transport conditions and resulting quality development • Being able to have insight into the distribution of trays over retail outlets • Tracking of pallets (or other returnable packaging materials) for finished goods

Generally speaking, trust is 'a psychological state comprising the intention to accept vulnerability based upon positive expectations of the intentions or behaviour of another' (Rousseau *et al.*, 1998: 395). The successful generation of trust needs time and voluntary reciprocal interactions (Schweer and Thies, 2003). Trust is considered pivotal for the emergence and maintenance of cooperation in society, including the development of business relations and networks (Van Dijk *et al.*, 2003; Zucker, 1986). Shapiro *et al.* (1992) describe the development of trust in business relationships in three stages that can be achieved over time. The first level is calculus or deterrence-based trust meaning that the decision to trust is determined primarily by a calculation of costs and benefits or the presence of penalties in the case of violated trust. The second level of trust building is gaining knowledge about and experience with the supply chain partner in order to make the partner's behaviour predictable. This level of trust – so-called knowledge-based trust – is founded on information instead of deterrence and control. The third level of trust – identification-based trust – stems from common intentions, values and norms. It requires reciprocal implementation of identical philosophies and convictions and trust as a general principle of organisation.

What exactly is the relationship between trust and transparency? On the one hand, trust can improve transparency. The general comprehension of transparency, i.e. easy and rapid information exchange, implies the willingness of supply chain partners to cooperate. If supply chain partners accept vulnerability and are willing to take the risk that the partner could possibly exploit the information provided (i.e. trust each other), then there will be more information exchange and, thus, a higher degree of transparency, for instance through improved traceability. But Hofstede (2003b: 4) also points out that 'people who trust one another do not need to tell each other anything just to show what they did. They only tell one another what they know the other needs to know.' In this sense, the demand for more transparency, i.e. the lack of willingness to accept an incomplete status of information, indicates a lack of trust in supply chain partners (Apelt, 1999).

Research into the determinants of trust has revealed that the voluntariness and mutuality of an action or a communication process are dominant factors in trust building because to show the intention to share information or to provide confidential data means to trust the partner. Therefore, if 'transparency is imposed from outside the Netchain, or if the Netchain is in fact institutionally a hierarchy and the leader imposes transparency', transparency and trust contradict each other (Hofstede, 2003b: 8). As a consequence, traceability (or, in Hofstede's (2003a) terms, history transparency) which is often imposed by the government or dominant supply chain partners such as retailers does not contribute to the generation of trust in food supply chains due to its obligatory character. Instead, tracking and tracing systems might be interpreted as signals of distrust because otherwise there would be no need to track or trace.

In contrast, transparency with regard to day-to-day business or strategic intentions (or, as Hofstede (2003a) put it: operations and strategy transparency) is valuable for trust building in supply chains because of its generally voluntary character. These forms of transparency are closely related to knowledge-based and identification-based forms of trust (Shapiro *et al.*, 1992). They, thus, lead to cooperative behaviour and effective information exchange among organisations (Jones and George, 1998).

In summary, we can say that there is a circular nature of the relationship between trust and transparency. On the one hand, trust improves transparency of food supply chains since it increases the willingness to share critical information (Hofstede *et al.*, 2004). On the other hand, transparency is a precondition and a mediator for trust (DiPiazza and Eccles, 2002) because information exchange – especially on a voluntary basis – includes a power-balancing and trust-building function. Due to this circular nature of the relationship between trust and transparency, both constructs should be enhanced at the same time.

In the following, two case examples will be given. The first is on the challenges for a Dutch organic pork chain to achieve traceability, the second case study will present a supplier evaluation system in the feed sector.

The Dutch organic pork meat supply chain

Although the market share of organic pork meat is still rather small in the Netherlands, it is gradually growing. The organic food market is one of the fastest growing markets in Europe, although the growth has slowed down in recent years. Denmark, Austria and Switzerland are examples where the market share is high in comparison with the Netherlands, Belgium, France and Norway (Padel *et al.*, 2009; Schaak, 2009). Within Denmark, organic food reached a market share of 7.2% in 2009, while in many European countries, the threshold of 2% market share has not been passed (Willer, 2009), including the Netherlands where the market share in 2010 was 1.7% for organic products and 1.9% for organic pork meat.

Within the Dutch organic pork meat supply chain, it is not yet possible to communicate information on individual farms to consumers when they buy a package of meat. Production information is made available through the package label and the website of De Groene Weg (www.degroeneweg.nl). Besides meat type, expiry date, weight and price, other information elements that can be found on the package label are: the nutritional value, the brand, the country of origin and the specific place where the pig is slaughtered and processed. However, full traceability of a piece of pork is not yet possible for consumers. For instance, numbers are referring to both slaughterhouse and processing line without indicating what those numbers stand for. Such information is only meant to comply with legislative requirements.

Another possibility to gain extra information is offered through the website of De Groene Weg where information can be found on characteristics and advantages of organic processes and of a selection of organic pork farmers. Information on farms with additional pictures of both farm and founder(s) has been included. Farmers may explain why they have started or converted to organic production, how animals live and how they manage their farm. However, this farm information is not (yet) linked to a package of fresh organic pork meat.

One bottleneck that should be overcome to achieve traceability from consumer to farmer is the loss of farm identification during the slaughter process, which means that it is not possible to link an individual farm to the organic pork cutlet bought at the retailer. Currently, after slaughtering the pigs, the carcasses are cleaned, examined, and cut in two halves. In addition, the quality is checked and rated, in particular the meat-fat percentage. The two carcass parts are hung on a slaughter hook. The carcass parts are stored in a cooling storage for 24 hours sorted according to the quality levels that have been identified. In the cooling storage, carcasses from different farmers are mixed. After cooling, carcasses are cut into six meat parts resulting in three different types of meat parts. These meat parts are hung type by type on meat hooks. The meat parts are transferred on the meat hooks to the processor for further cutting and processing. All meat parts together are treated as a day batch.

Within the Dutch organic pork supply chain, an individual identification number (including the farm identification number, called UBN) is attached as a plastic ear label to a piglet on the farm. Before delivery to the slaughterhouse, an additional metal individual identification number with the UBN and a sequential number is attached to the pig ear because the original label often gets lost or damaged during the slaughter process (e.g. in the burner to clean the pig skin). Within the slaughter process, six stamps with farm-identification number are printed on the carcass which means that six cuts of the carcass can be delivered to the processor without losing the UBN. However, this number is currently not used because the decision has been made to consider meat from pigs slaughtered on one day as belonging to one batch while a batch consists of pigs from several farms. Further use of these UBNs for identification purposes would mean extra investments in data capturing, storage and processing capabilities.

Another bottleneck for instant information delivery towards the end consumer is the lack of interoperability between information systems within the Dutch organic pork meat supply chain. If farmers are to be traced from a package of meat, information from several information systems needs to be retrieved and matched which cannot yet be done automatically.

To remove these bottlenecks, unique identification should be extended to meat parts and even end-products, e.g. by RFID technology. Moreover, integration of information systems of the different actors in the chain should be implemented.

The Allianz Futtermittel Sicherheit – supplier rating system

The Allianz Futtermittel Sicherheit (AFS) Deutschland eG was founded in the autumn of 2011 against the background of a dioxin crisis in Germany. The compound feed industry did not cause the crisis but is expected to prevent the next. Therefore, the DRV (Deutscher Raiffeisenverband e.V.) and its members started searching for approaches for a supplier rating system in order to improve the quality of raw materials and to prevent that undesirable substances get into the feed.

AFS was modelled on the basis of TRUST FEED (www.trustfeed.eu) which was established in the Netherlands in 2011 from the former TrusQ and SAFE FEED. There is a close collaboration with the TRUST FEED, but no data exchange exists yet. The AFS database has been working since 2012. At the beginning of 2013, the AFS had 16 members producing nearly two million tonnes of compound feed. 525 suppliers with 255 different products are listed.

The database contains information about the company (members), their suppliers, the status of the certificate, the products and their related risk categories, feed safety datasheets, the audits (results, deviation list, arrangements, time limits and supervision), rejected freights and complaints. The database is a supporting tool for planning and conducting the post-processing of audits. The processing and documentation of corrective actions is another core element. The information about audit results is exchanged. In addition, analytical results and reports are provided for the members. The supplier rating data are transparent. The system enhances traceability and trust. Figure 6.3 shows how AFS works.

AFS is an active tool for the prevention of crises through:

- daily updated information on the status of GMP/QS certification of own suppliers;
- systematic administration of data from suppliers and raw material for each member;
- provision of product safety datasheets of listed suppliers;
- recording and evaluation of rejected freights and complaints;
- rapid-warning system about rejected freights;
- in case of problems, immediate information from affected members as relation between supplier, product and member is available via the AFS database;
- provision of evaluable audit reports and analyses of contaminants.

Thereby it enhances traceability in the feed chain and gains trust in the whole meat chain. To comply with the QS system (www.q-s.de) and the GMP+ system (www.gmpplus.org), animal feed producers are already obliged to assess their suppliers. AFS offers an enhanced way to evaluate suppliers. It amends these systems and improves transparency for and trust in supply chain partners.

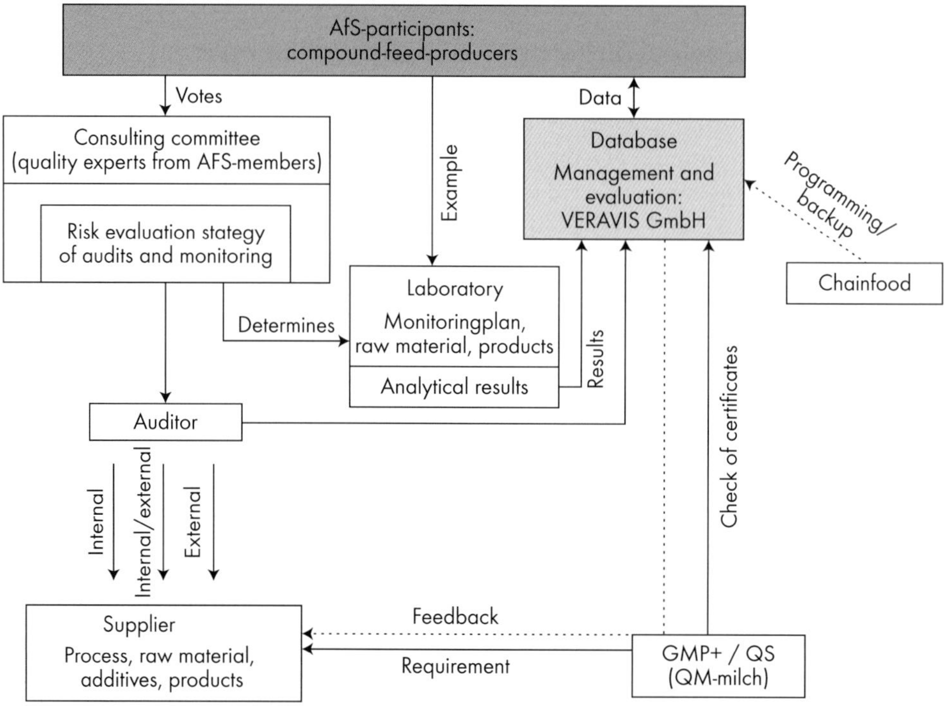

Figure 6.3. Organisation and information flow in the Allianz Futtermittel Sicherheit (AFS; http://afs.chainfood.com/afs-struktur.php).

6.3 Innovation management in agri-food networks & clusters

M. Bruns, B. Hundt and M. Hamer
GIQS e.V., c/o University of Bonn, Katzenburgweg 7-9, 53115 Bonn, Germany
m.bruns@giqs.org

Within a modern industrial economy, there is an enormous need for continuous innovation in order to compete with or to have a lead over competitors, independent of the sector. This applies both on the micro and macro-economic level. All companies and all sectors are confronted with increased competition resulting from the open internal European and global markets. To meet this competition, scientific research, technological development and innovation are crucial. They represent core aspects of the knowledge-based economy. On the micro-economic level, innovation is of relevance since the competitiveness of single companies depends on their innovative capability. The competitive pressure caused by globalisation requires continuous improvement of performance. Furthermore, increasing competition generally leads to shorter product life and innovation cycles (Vahs and Burmester, 2005). At the macro-economic level, innovation is of great importance since it is associated with large investments (e.g. construction of research facilities, acquisition of operating resources or recruitment of additional personal). Besides that, Vahs and Burmester (2005) have detailed the interrelation between these investments and a positive effect on turnover and acquisition activities. Innovation is combined with multiplier and capital accumulation effects. Due to this, innovation becomes the driving force behind economic development. A positive relationship between innovation activities and annual turnover can be observed also in the food industry (Figure 6.4, which is based on a German example).

The best way to exploit the potential of different innovation activities is through structured planning and control of innovation activities from the initial idea to successful market entry. Based on experiences in innovation consulting, A.T. Kearney developed the 'House of Innovation' (European Communities, 2008). In Figure 6.5, the slightly modified model of successful innovation management including important elements is shown (Bruns, 2011).

Single and multi-actor innovation

Innovations can be implemented in-house (by a single enterprise) if the enterprise has the necessary resources and competences available internally and if the topic of the innovation activity concerns only this single company (e.g. food product development). Another strategy for increasing innovation levels is to use external resources (Chesbrough, 2003). In this case, an enterprise makes use of an innovation system which involves the interaction

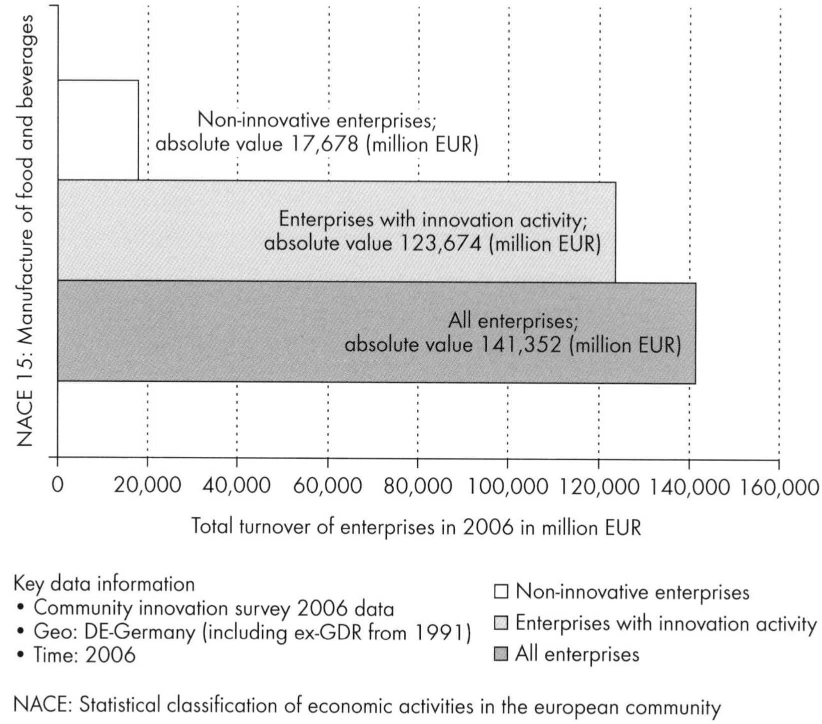

Figure 6.4. Relationship between the implementation of innovation activities and annual turnover in the food industry (based on data from the 2006 Community Innovation Survey – CIS).

between actors that is needed in order to turn an idea into something new that is to be introduced on the market. This is often necessary if the topic of the innovation activity has a value chain or a value network perspective. These are innovations with an impact on the organisation of these systems. In this event, more than one actor is needed (Figure 6.6).

The more complex the institutional system working on an innovation (from one enterprise to a whole network of actors) is, the more organisations and actors need to be involved in the innovation activity. If more than one actor is involved, it can be called a multi-actor innovation. If the entire system (chain and network perspective) including all its public and private stakeholders needs to be developed further, the innovation has to be based on a multi-actor innovation (for example the development of a new image or a new sustainable production system standard for all companies in a certain region). Innovations are mainly results of a complex set of relationships among actors like enterprises, public authorities, universities and research institutes (Boon, 2008; Freeman, 2002; Klerkx, 2008; Lundvall, 1992). The term 'open innovation' describes the approach of companies that open up

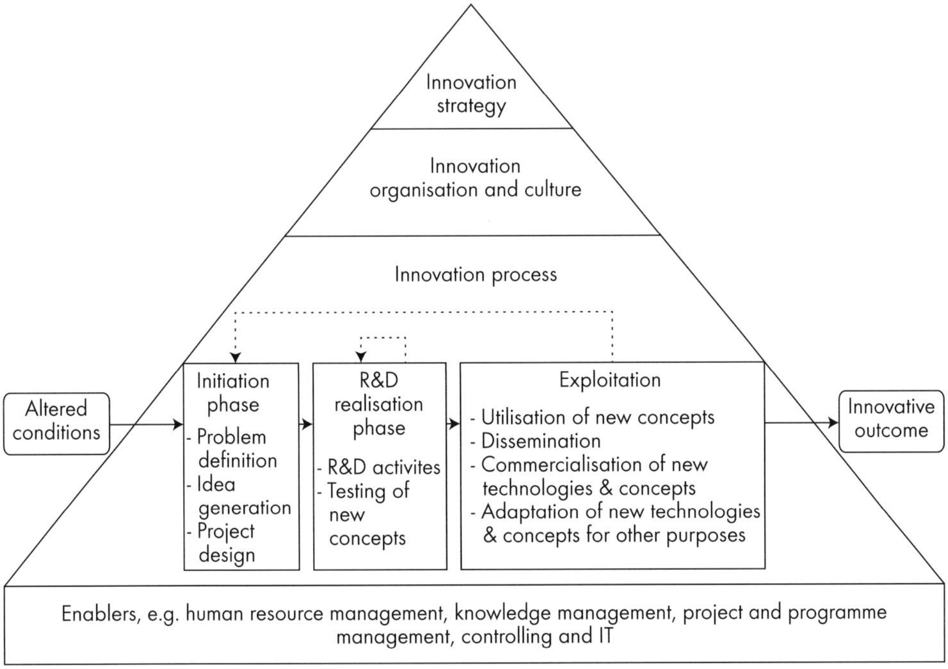

Figure 6.5. 'House of Innovation' (modified after European Communities, 2008; furthermore based on Bruns *et al.*, 2008; Menrad, 2004; Rosenfeld and Servo, 1991; Rothwell, 1994; Schlicksupp, 1992; Schroeder *et al.*, 1986; Trommsdorff, 1990).

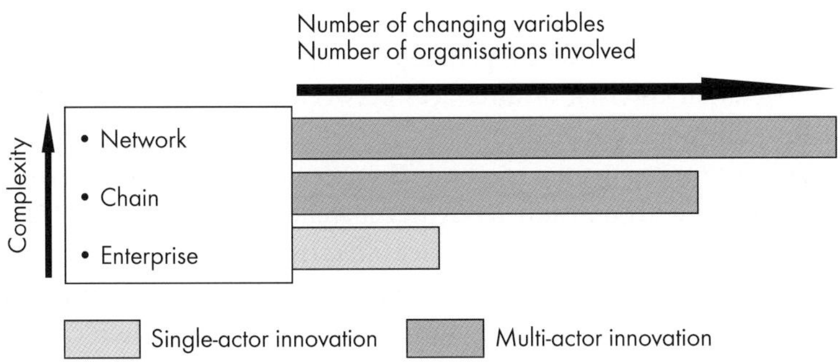

Figure 6.6. Scope of innovation regarding the number of organisations involved (modified after Bruns *et al.*, 2008; NRLO, 1999).

their institutional boundaries for other actors to implement innovation activities with the aim of stimulating innovation instead of solely innovating internally. Social sub-units can establish an innovation system in order to bring forth innovations. These are called innovation networks or clusters. The interlocking of research institutions, businesses and public actors, including the functioning transfer of technology, is a key to strengthening innovation power and thus a motor for the growth process (Buhl, 2009). It is a matter of innovation systems on a small scale with similar system characteristics and interactions (Den Hertog and Roelandt, 1999).

In a cooperative development, costs and risks can be shared. As well as this, a considerable advantage for participating companies is the keeping up to date with the latest technological developments and to have access to new technologies and knowledge. Innovation through collaboration with competitors, suppliers and research institutions also gives new insights into the value chain. Gemünden *et al.* (1996) illustrated that innovation success is significantly correlated with a firm's network. However, the complexity of inter-organisational innovation projects could make it difficult for companies to participate.

Barriers that hamper inter-organisational innovation activities

Some innovation barriers related to the implementation of innovation processes mentioned in literature (Aslesen *et al.*, 1999; Batterink *et al.*, 2006; Bruns, 2011; Buhl, 2009; Fortuin *et al.*, 2007; Henry and Walker, 1991; Klerkx, 2008; Klerkx and Leeuwis, 2008; Lienemann and Lehnert, 2005; Rammer *et al.*, 2006) can be overcome by inter-organisational innovation processes. But it is difficult for a large number of enterprises to initiate and participate in innovation cooperations. A survey conducted by Lienemann and Lehnert (2005) indicates several barriers in connection with R&D cooperation expressed by SMEs in the German agri-food sector (Figure 6.7).

In relation to potential innovation cooperations, companies as well as research institutions are complaining about the lack of interfaces between business and science resulting in an obstacle for innovation cooperations (Aslesen *et al.*, 1999; Lienemann and Lehnert, 2005). It is difficult for companies to obtain an overview of the research landscape, and so the knowledge market is not sufficiently transparent (Aslesen *et al.*, 1999; Buhl, 2009; DGQ, 2000). In addition, the quality of service offers from cooperation partners is difficult to assess beforehand (Czarnitki *et al.*, 2001; Klerkx and Leeuwis, 2008; Lienemann and Lehnert, 2005; Pollard, 2006). In this context, the notion of reputation can also be cited. The quality of cooperation partners can often only be rated through good experience or through their reputation. Hereby trust in the competence of potential cooperation partners plays a decisive role. Scepticism and lack of trust towards cooperation partners can impede the readiness to cooperate (Buhl, 2009; Czarnitzki *et al.*, 2001; Lienemann and Lehnert, 2005; Pollard, 2006). Furthermore, actors within a cooperation have to come to terms with different work cultures: while companies follow the generation of quick solutions, the research institutions concentrate more on the applied method. This can lead to business and science working in different chronological dimensions (Klerkx, 2008). Limited by the large

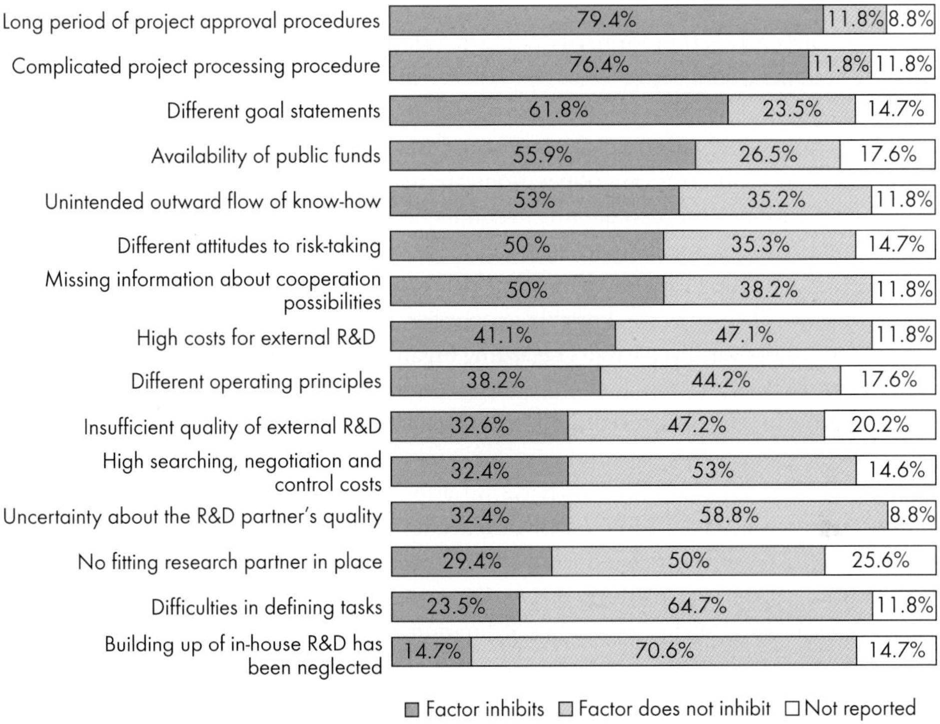

Figure 6.7. Barriers in connection with R&D cooperation in the agri-food sector (Lienemann and Lehnert, 2005).

circle of actors, the coordinated decision making mechanisms must be implemented and complied with which can, however, slow down the coordination process (Buhl, 2009). In accordance with this, longer process running times present further barriers to the innovation cooperation process because of the comprehensive coordination of the actors involved.

A very significant aspect which is of great importance to the innovation cooperation is the subject of intellectual property. Even on the individual business level and especially in SMEs, a lack of knowledge and high searching, application and policing costs can be observed in this area. If innovation is generated within a cooperation, there can be problems with the perception of ownership, i.e. what counts as individual business knowledge and what counts as joint knowledge has not been clearly defined (Ladeur and Vesting, 2008), for example. Further challenges (possible boundaries for networks of individual partners) would be: no sole use of patents of newly developed product/process innovations, or also the exposure of personal knowledge when opening up to competition (Buhl, 2009). Regulations for handling joint intellectual property as well as already existing knowledge that individual partners bring should definitely be laid down in contract. Such contractual regulations offer protection for network partners and can support the necessary openness within the cooperation.

Barriers in connection with innovation processes that have an impact on the initiation and implementation of innovation cooperations are relevant for all companies, even if these barriers are mentioned more frequently in connection with SMEs. It can be assumed that in cooperation situations involving scientific and industrial actors, special problems regarding planning (e.g. searching for appropriate partners) and management (e.g. the coordination of consortia) are likely to occur. Possible reasons for these difficulties might be the different backgrounds of the various partners regarding organisational culture and more specifically research approaches, the availability and deployment of missing resources, and, not least, the huge administrative effort required for the management of innovation networks.

In any case, barriers in connection with the initiation and realisation of inter-organisational innovation processes need to be taken into account. Barriers occur due to the fact that actors need to make efforts to handle uncertainties by opening up their institutional borders (Williamson, 1985). These efforts can be summarised under the term 'transaction and coordination tasks'. Transaction and coordination failures accrue especially at interfaces between participating actors.

Reducing transaction and coordination barriers by integrating a third party

The above mentioned failures might be minimised by a third party taking over coordinating tasks. When integrating a third party into inter-organisational innovation activities, this party should focus on the reduction of transaction and coordination barriers. 'Such intermediary activities include: helping to provide information about potential collaborators; brokering a transaction between two or more parties; acting as a mediator, or go-between, bodies or organisations that are already collaborating; and helping to find advice, funding and support for the innovation outcomes of such collaborations' (Howells, 2006). According to Howells (2006), this third party might be 'an organisation or body that acts [as] an agent or broker in any aspect of the innovation process between two or more' actors. Winch and Courtney (2007) define the term 'innovation broker' as 'an organisation acting as a member of a network of actors in an industrial sector that is focused neither on the generation nor the implementation of innovations but on enabling other organisations to innovate.' The innovation broker does not participate directly in the generation of new knowledge during the innovation process. Assuming, by the integration of a third party supporting another organisation to innovate, inter-organisational innovation consortia would be more efficient and effective since the actors within the consortium would be able to concentrate solely on the content of an innovation activity – the generation, adaptation and exploitation of new knowledge.

According to Bruns (2011), the scope of support for companies integrated in an inter-organisational innovation process grows considerably depending on the following five factors:

- the size of the cooperation consortium;
- the lack of appropriate personnel with educational backgrounds and experience in the area of innovation / R&D;
- lack of knowledge and no access to the new target markets being aimed at (based on the markets of the agri-food industry);
- lack of strategic alliances between steps in the value chain;
- lack of experience in initiating, applying for and implementing publicly funded projects.

Quality aspects in networks & clusters

The role of the innovation broker is not only in demand in the case of market failures or failure of the innovation system. Innovation brokers offer important innovation management support services (Table 6.3) within the framework of innovation networks. These support services can be seen as a kind of outsourcing of innovation management tasks from the point of view of network members (Bruns, 2011).

Also to be mentioned is the performance of offering support services influencing the quality of innovation network management. According to that, the European Cluster Excellence

Table 6.3. Catalogue of management support service elements provided in inter-organisational innovation processes (based on Bruns, 2011).

	Innovation management support services (a selection)
Networking	• Bringing together actors of the innovation system • Lobbying (e.g. funding programme setting)
Initiation phase	• Identification and definition of innovation demand • Idea generation and evaluation (including mirroring with external market information) • Resource management: identification of needed and internally available resources, procurement of external resources / know-how • Product design development and planning (objectives, milestones, responsibilities, timescale, budget) • Elaboration of the project budget (including request of public funding) • Setting up a consortium agreement, if needed
Implementation phase	• Establish processes and communication structures between project parties involved • Project plan monitoring and controlling • Implementation of new technologies, products and concepts into day-to-day management (change) • Dissemination of innovations to staff and public (PR, trainings)

Initiative (ECEI) has developed a set of quality indicators to measure and improve cluster management excellence (Table 6.4).

Table 6.4. A selection of quality indicators developed within the framework of the European Cluster Excellence Initiative (based on Meier zu Köcker, unpublished data).

Dimension	Indicator
Structure of the cluster	• Committed cluster participation • Composition of the cluster participants
Typology, governance, cooperation	• Maturity of cluster management • Human resources available for cluster management • Qualification of the cluster management team • Life-long learning aspects for the cluster management team • Stability and continuity of human resources of the cluster management team • Stability of cluster participation • Involvement of stakeholders in decision-making processes • Direct personal contacts between the cluster organisation management team and cluster participants
Financing	• Prospects of the financial resources of the organisational cluster • Share of financial resources from private sources
Strategy, objectives, services	• Strategy building process • Documentation and review of cluster strategy and implementation plan • Activities and services of cluster management (incl. performance monitoring) • Financial controlling system • Communication activities of cluster organisation (PR, etc.)
Achievement, recognition	• Recognition of the cluster in publications, press, media • Success stories • Customer and cluster participants' satisfaction assessment

6.4 Cold chain management: new packaging concepts to reduce food waste

J. Kreyenschmidt, U. Herbert, S. Rossaint and Y. Ilg
University of Bonn, Cold Chain Management Group, Katzenburgweg 7-9, 53115 Bonn, Germany
j.kreyenschmidt@uni-bonn.de

Introduction

The most important criteria from a consumer and industry perspective are high quality and safe products. From the industry point of view, long shelf life times are important for economic reasons. At the same time, the competition within the field of perishable food chains is steadily increasing and sustainability aspects are gaining in importance. Therefore, cold-chain management is of vital interest as it ensures high product quality and safety of perishable products. Furthermore, an effective cold-chain management optimises inter and intra-company logistical processes, resulting in a reduction of food waste.

Most of the perishable food products provide good bacterial growth conditions due to their nutritional composition and physical properties. This often results in short shelf life and selling times which leads to an increased risk for food waste. In the meat sector, for example, the rate of wasted products from production to consumption is assumed to vary between 12-22% in Europe (Kreyenschmidt *et al.,* 2013).

One important reason for wasting food products is wrong handling so that the product is spoiled before the use-by-date is reached. For example, the shelf life of fresh pork is reduced by 30% if the average storage temperature is increased to 4 °C instead of 2 °C. Also short time temperature interruptions have a significant influence on the shelf life of perishable products. Shelf life reductions of up to 20% were observed for fresh poultry and fresh pork when the periods with out-of-range temperatures were less than 5% of the total storage time and the temperature shift was 11 °C (Bruckner *et al.*, 2013).

Besides that, several food products are wasted at the retailer stage before the best-before or use-by-date is reached even if the products are still in good condition. Retailers are aware of critical food safety or quality parameters because effective monitoring strategies over the chain are missing. Also several products are thrown away because they have not been sold during the short selling time.

The application of novel packaging strategies can make an important contribution to reduce the amount of food waste in perishable food chains and to ensure sustainable food production. They increase the shelf life and the selling time or provide additional information on the real product quality, safety and history of the product over the entire supply chain. Figure 6.8 illustrates different approaches to optimise food packaging.

In the following, different packaging strategies to reduce food waste in perishable food chains will be demonstrated. The first part will deal with new developments in the field of modified atmosphere packaging (MAP) with regard to shelf life and sustainability aspects. The next part will focus on active packaging solutions to prolong shelf life. In the following, intelligent packaging solutions resp. labels which deliver information on the product history, quality and remaining shelf life will be illustrated.

Modified atmosphere packaging and new developments

The gas atmosphere inside the packaging is in addition to the temperature one of the most important influence factors on shelf life. The so called Modified Atmosphere Packaging (MAP) was introduced in the market in 1979 at the retailer Marks and Spencer (Church, 1994). It is a preservation technique where the surrounding air of a product is replaced by a selected mixture of gases. MAP is commonly established, and the trend of self-service merchandising of fresh meat became more and more important during the last decade (Bell, 2005). Approximately 50% of the meat products were purchased in self-service by private households (www.ami-informiert.de/ami-english.html). The commonly used gases for perishable products are oxygen (O_2), carbon dioxide (CO_2) and nitrogen (N_2) which were added in the package headspace in different proportions and combinations (Phillips, 1996; Floros and Matsos, 2005). The shelf life of fresh poultry, for example, can be prolonged by 25% (4 °C) if poultry is packed under modified atmosphere instead of aerobe conditions (Figure 6.9).

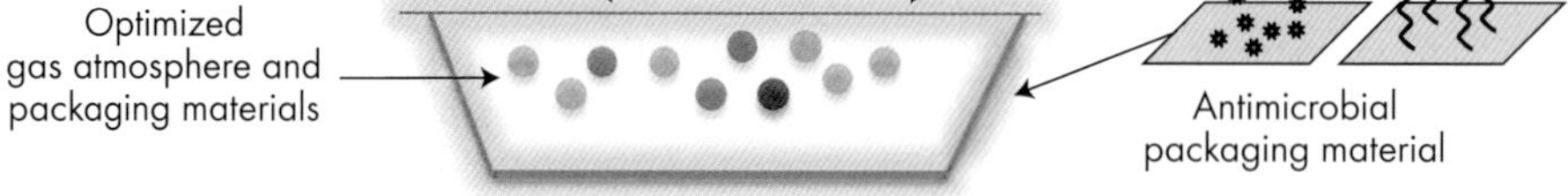

Figure 6.8. Different packaging strategies to reduce food waste.

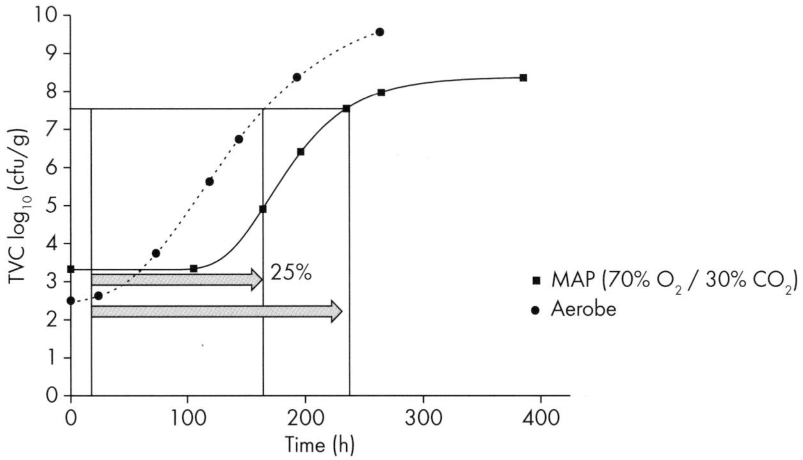

Figure 6.9. Effect of modified atmosphere on shelf life extension of poultry meat.

Even though the technology has been introduced for several years now, there are still discussions about the optimum gas atmosphere. Also the application of new gases like argon, helium or carbon monoxide with regard to prolongation of shelf life are discussed. Several studies were performed to investigate the effect of argon on the shelf life of different meat and meat products (Fraqueza and Barreto, 2009; Herbert *et al.,* 2013; Ruiz-Capillas and Jiménez-Colmenero, 2010; Tománková *et al.,* 2012). No significant effects on shelf life extension could be observed using argon in a packaging atmosphere. However, some studies reported a beneficial effect on sensory attributes (Herbert *et al.,* 2013; Ruiz-Capillas and Jiménez-Colmenero, 2010).

With a view on sustainable packaging concepts, the use of MAP is discussed since a headspace:product ratio of 3:1 is not ideal with regard to environmental aspects. Decreasing the ratio to 2:1 leads to a reduction in sensory shelf life of approx. 25% (Rossaint *et al.,* 2013, unpublished data). Thus, further research is necessary to balance the packaging size, the thickness of packaging material and the length of shelf life. During the last years, the application of bio-based polymers has more and more been discussed in order to reduce the amount of valuable primary resources for packaging material. Polylactic acid (PLA) is the most common one produced synthetically from renewable corn or whey. For MAP application, biomaterials have to be adapted to the properties of the product. For the application in the meat sector, for example, sealing requirements or gas permeability have to be further improved in the next years (Baroni *et al.,* 2009; Holm *et al.,* 2006).

To overcome these aspects, another new packaging concept was developed, the so called vacuum-skin packages. It is a packaging technique where the product is placed in trays; a polymer film is shrunken around a product using heat and the film adheres to the tray and product. This results in a skin-tight pack (Lagerstedt *et al.,* 2011). The shelf life of fresh meat for example is similar to the one of MAP but the lack of the bright red colour compared to MAP is often not accepted by the consumer (Taylor *et al.,* 1990). By using this technology, however, the packaging material, the volume and thus the logistic and storage costs will significantly be reduced. Such a packaging approach is attractive for the producers, logistic companies, wholesalers and retailers with regard to sustainable food production (Vázquez *et al.,* 2004). Further beneficial effects in the field of packed products can be reached by combining active and intelligent packaging strategies with MAP.

Active packaging

Another approach to actively prolong the shelf life and/or to improve food quality or safety of perishable products is the active packaging strategy. It describes the integration of devices or antimicrobial components in the package which interacts with the product itself or its environment (Coma, 2008; Ilg, 2010; Simon *et al.,* 2008).

Scavengers are packaging films or devices that can be applied in MA or vacuum-skin packaging. They react with the gas within the packaging to remove it from the food environment. Examples are oxygen or ethylene scavengers. Oxygen scavengers are binding the remaining oxygen within the packaging leading for example to a reduced growth of aerobic bacteria; whereas ethylene scavengers are capable to reduce the ripening of fruits and vegetables (Vermeiren *et al.,* 1999).

Antimicrobial components actively inhibit the growth of bacteria or even kill them during food storage. Most of these agents are based on surface release, i.e. a migration of antimicrobial agents into the food product is required. To generate this effect, a direct contact between the packaging material and the food product is essential as is the case in shrink, skin or vacuum packaging (Cooksey, 2001). Antimicrobial active components can also be integrated in soaker pads or trays which are used, for example, in modified atmosphere packaging technology.

In recent years, various migrating antimicrobial components have been investigated and developed. They range from organic acids and biocides up to plant extracts or (nano-scaled) metal compounds. Reduction in bacterial count in the area of log 1.0 cfu/g can result in differences in shelf life of several days. Lee *et al.* (2004) reported a significant decrease in bacterial growth in milk by adding nisin and/or chitosan to the inner layer of the package. Also the number of spoilage organisms in beef was reduced by nisin-activated packages (Ercolini *et al.,* 2010). The integration of lysozyme into surfaces is described by Padgett *et al.* (1998), resulting in a good antimicrobial activity against *Lactobacillus* and *Escherichia*. Brody *et al.* (2001) relate that crepe paper treated with sorbic acid was effective against fungal growth in bread. Antimicrobial metal compounds are the most common migrating antimicrobial additives in food packaging due to their temperature and mechanical stability. Kampmann *et al.* (2008) describes that food stored on silver-coated surfaces results in a better quality compared to storage on uncoated surfaces.

However, it has to be considered that the activity of several migrating antimicrobial components common in perishable food chains (Kampmann *et al.*, 2008) is reduced at low temperatures. Also proteins and some other food residues may suppress the antimicrobial activity. Thus, for example most silver-containing packaging materials are not usable for meat and other food products rich in proteins (Ilg and Kreyenschmidt, 2011). Furthermore, migrating antimicrobial compounds may be released into the environment.

A new class of bioactive polymers are sustainable active microbicidal (SAM) polymers (B. Kossmann, personal communication). In contrast to migrating components, the active part is fixed on the surface here. Because of its low toxicological level towards mammals, SAM polymers are in general suitable for the use in food packaging (Buranasombop, 2005). However, the mechanical and technical properties of these polymers need further development for their successful application as bactericidal food contact surfaces (Ilg *et al.*, 2011).

Intelligent packaging

Intelligent packaging is delivering information about the history, quality, safety or the shelf life of a product to every participant in the chain. The information is illustrated by small labels which are changing in colour and which are integrated in the package. During the past few years, several different types of labels have been developed.

Amongst the first intelligent labels to control the temperature conditions over the entire supply chain were the so called time-temperature indicators (TTI). These are devices that integrate the complete time-temperature history of their next surroundings. The time-temperature history is visualised by a colour change. Several different systems are available on the market. The principle of different labels is based on different reactions, like enzymatic, chemical, mechanical, electrochemical or microbiological reactions (Taoukis, 2008).

In comparison to other temperature monitoring strategies, TTIs allow monitoring of the temperature history on a single item level and thus from packaging to consumption. If the colour change of the label correlates with the spoilage kinetic of the food to which it is attached, the label can also be used as Freshness-Indicator. It will deliver more precise information about the remaining shelf live in addition to the use-by date or best-before date.

Furthermore, typical weak points in the chain like temperature abuses during the loading and unloading phase or temperature variations in the truck can be detected at an early stage, and the cold chain condition will be improved in the long term. Figure 6.10 shows different TTI systems.

Another type of label in the field of intelligent packaging is represented by Freshness Indicators. In contrast to TTIs, these labels are measuring the real quality. The labels are showing, for example, the accumulation of metabolic by-products of microorganisms (ethanol, acidic acid, CO_2, amines) or changes in the pH value of the product which are directly correlating with the freshness of the products. Other labels are showing the presence of pathogenic bacteria. The principle is based on an immunological-chemical reaction. If pathogenic bacteria are present, the bar code is not readable anymore. Further intelligent labels show if oxygen is present or not, while other types of label indicate the time passed after opening a package.

But up to now, a broad range of applications of intelligent labels in cold chains is still missing. The reasons for this are as follows: the food industry is not willing to implement such labels because one is afraid of high losses due to sorting the products by consumers within the retailers' shelves and one is also concerned about what will happen after the point of sale. Furthermore, the industry is critical of the beneficial effect of TTI on sustainable food production and is also afraid of the implementation costs.

Conclusion

Overall, new packaging strategies deliver an important contribution to the extension of shelf life, selling times and the reduction of packaging material. They also allow the monitoring of critical food quality and safety parameters throughout the food supply chain.

The implementation of modified atmosphere packaging several years ago was an important step in prolonging the shelf life of food products. The new developments in the application of Active Packaging like antimicrobial packaging material will support the effect of MAP and vacuum-skin packaging. They will provide a further beneficial effect on the length of shelf life and thus on the selling time. This supports the reduction of waste of perishable food.

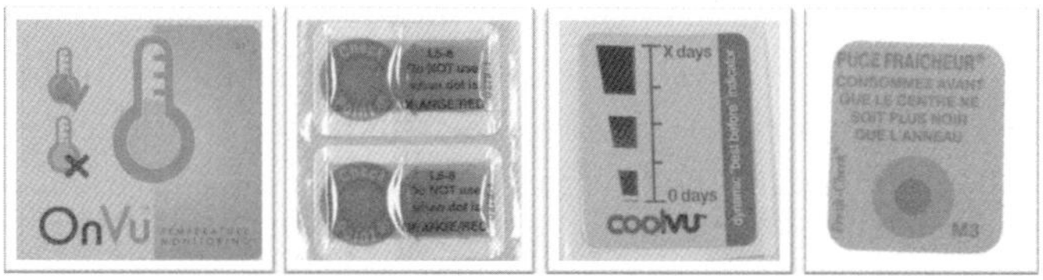

Figure 6.10. Different types of time-temperature indicator.

Furthermore, the integration of intelligent packaging solutions will deliver information about the product history such as temperature conditions or the real quality by showing the accumulation of metabolic by-products of microorganisms. By having real time information on typical quality and safety parameters in each step of the chain, wasting of products still in good condition can be significantly reduced. Also the storage management can be improved in a more quality-oriented way instead of using the 'first in – first out' principle.

To achieve the described positive effects of the new packaging strategies on foods waste reduction, open communication between all actors within the food chain is required. In addition to implementation, the balance between sustainability aspects – including food waste reduction – and economic aspects has to be considered. To assess new packaging concepts with regard to the mentioned factors, an online tool was developed within the framework of the EU IQ-Freshlabel project (FP7-243423). This tool enables calculating the cost and benefit for the implementation of different packaging concepts and the effect on sustainability and economic factors at different steps of the supply chain (Rossaint and Kreyenschmidt, 2013).

6.5 Supply chain environmental management

K. Grekova[1], J. Trienekens[1], H. Bremmers[2], L. Heres[3] and S.W.F. Omta[1]
[1]Wageningen University, Social Sciences Department, Management Studies Group, Hollandseweg 1, 6706 KN Wageningen, the Netherlands
[2]Wageningen University, Social Sciences Department, Law and Governance Group, Hollandseweg 1, 6706 KN Wageningen, the Netherlands
[3]VION Food Group, Noord Brabantlaan 303-307, 5657 GB Eindhoven, the Netherlands
ekaterina.grekova@wur.nl

Apart from quality and safety of food and animal welfare, the environmental impact of the food industry is increasingly capturing the attention of public authorities, NGOs, consumers and local communities. This is not only due to the intensive usage of natural resources during the food production but also because other issues in the industry (such as animal welfare including humane slaughter, fair trade especially in international supply chains, food safety scandals) alert the consumers and increase the burden on food supply chains (Maloni and Brown, 2006). Despite the pressure to reduce environmental impacts, in the past, many firms were reluctant to invest in environmental management because the benefits of environmental management were uncertain and they were afraid that these could provide positive innovation spill-overs for other firms and benefit the whole society at the cost of individual firms (Rennings, 2000). In the past few years, the perception of environmental management has changed: from considering environmental management as a source of additional costs to viewing it as a possibility for cost savings (e.g. through the reduction of energy, water usage, waste treatment costs, etc.). The potential of environmental management to benefit the environment and saving costs provides an incentive to food firms to invest in environmental practices.

The food industry is one of the largest industries in EU (12.9% of household consumption is attributed to food (Eurostat, 2010)) and it keeps growing. Consequently, more pressure on food firms to reduce their waste streams, water and energy consumption, pollution, and CO_2 emissions is observed. In addition to environmental demands, the consumers' demand is focused on maintaining and increasing food quality and safety, while keeping the price low at the same time. Although the food industry improved its efficiency considerably over the past few years, the challenges of cost saving and increasing the efficiency remain. This is, amongst others, due to pressures downstream of the supply chain, retailers' competition and competition between the private labels of the retailers and brand manufacturers (Erich, 2012). In many developed countries, an intensive cooperation to increase efficiency is observed between the partners occupying different stages of supply chains: farmers/growers, processors, and wholesalers/distributors/retailers. As the sustainability concerns in the industry are becoming more important, supply chain cooperation increasingly addresses environmental issues.

Nowadays, firms are getting more experienced in environmental management and in-house environmental practices. It becomes clear, however, that environmental impact reduction targets cannot be achieved solely by the effort of individual firms in the chain: firms cannot mitigate the environmental impact of the output (final product) without taking environmental characteristics of inputs from suppliers into account. Also, cooperation with downstream chain partners is essential to optimise transportation, environmental product (re)design, reverse logistics, etc. Cooperation with up and downstream supply chain partners to reduce environmental impact has been studied intensively for about a decade using different terms such as supply chain environmental management (SCEM) (Sharfman *et al.*, 2009), 'green supply chain management' (Zhu *et al.*, 2008b), 'sustainable supply chain management' (Gold *et al.*, 2010), 'green purchasing' (Min and Galle, 2001), etc. SCEM is associated with better environmental performance and operational performance (Vachon and Klassen, 2008; Zhu *et al.*, 2005). SCEM is also related with product quality improvement (Azevedo *et al.*, 2011; Zhu *et al.*, 2010). Because environmental management requires more sophisticated and in-depth technical and managerial knowledge, environmental management can lead to a better understanding of business processes and, ultimately, can provide positive spill-overs for the product quality (Pil and Rothenberg, 2003). There is also evidence that SCEM contributes to business performance (see e.g. Azevedo *et al.,* 2011; Zhu *et al.,* 2010) especially in terms of cost savings (De Giovanni and Esposito Vinzi, 2012). Consumers might also reward more environmentally friendly products with a price premium. In summary, SCEM has the potential to provide superior environmental performance, operational (including quality) and business performance (including cost reduction). Moreover, collaboration in food supply chains is increasingly seen as a tool to improve their competitiveness (Manzini and Accorsi, 2012). The benefits promised by SCEM are crucial for food firms because they are challenged to meet major customers' demand for high quality products, sustainability and low price.

The coordination and content of supply chain environmental management practices in food supply chains

The specificity and complexity of food supply chains influence the coordination of SCEM efforts between supply chain actors. Food supply chains are typically characterised by the large amount of farmers/growers upstream of the supply chain and a relatively small number of large and powerful retailers downstream of the chain. Mostly retailers, large and powerful food service companies, and branched industry impose environmental requirements on food supply chains. For instance, Unilever committed to sourcing 100% of its agricultural raw materials sustainably by 2020 relying either on third party certification or on their own Unilever Sustainable Agriculture Code. To achieve sustainable supply in specific sectors, like dairy and meat, Unilever works with the large milk processor Friesland Campina and meat processor VION because they can cooperate better with many smaller supplying farmers to improve sustainability of their practices. Also, the largest Dutch retailer, Albert Heijn, stimulates SCEM in his supply chains by committing by 2015 to buy meat, dairy, and eggs coming from animals fed with RTRS[3] certified soy. Albert Heijn also tries to understand and map its complex supply chains to reduce environmental impact up to the last-stage production sites upstream of the supply chain. The pressure to reduce environmental impact is actively expressed by retailers. However, large food service companies, branched industry and food processors are often those who have to coordinate environmental management activities in supply chains. The coordinating role, the concentration and the power of food processors depends on the food sector. In some sectors, like meat and dairy, the processors are one of the major parties to coordinate environmental activities in supply chains. Taking into account the role of processors in supply chain coordination, our research is conducted from the perspective of food processors on SCEM.

SCEM typically refers to green purchasing, cooperation with customers including reverse logistics, and eco-design (Zhu *et al.*, 2008a). Green purchasing practices include eco-labelling, environmental cooperation, ISO 14001 certification of suppliers, suppliers' environmental audit, and evaluation of environmental practices of the second-tier suppliers. In food supply chains, suppliers' certification and especially audits by food processors are not realistic due to the often large number of suppliers and their small scale. Furthermore, the high expenses for certification might prove difficult given the generally small profit margins in the food industry and the small niche markets for premium-priced environmentally friendly products. Even if the farmers introduce environmental management systems able to be certified, based on the previous industry experience (e.g. IKB system), we can expect third parties rather than processors to conduct the audits. Limited opportunities to modify the product and material in the food industry also influence the content of SCEM in food supply chains. In many food sectors, the opportunity for eco-design refers only to product packaging and even the latter are rather limited. Especially reverse logistics practices are promising to improve environmental sustainability of food supply chains via recycling, reuse, and recovery of return products and packaging (Manzini and Accorsi, 2012).

Supply chain environmental management: determinants and performance implications: the case of meat supply chains

We collected data from 50 Dutch and German meat processors to illustrate the factors influencing SCEM development in meat chains and the impact of SCEM on business performance from the perspective of meat processors.

Table 6.5 indicates that there are more explicit reasons to introduce SCEM on the customer side than reasons to engage in supplier-side SCEM practices. This can be due to the fact

[3] RTRS stands for The Round Table on Responsible Soy Association which is a multi-stakeholder platform to develop a global solution to sustainable soy production and to prevent such environmental issues as deforestation, soil erosion, and pesticides usage.

that customer-side practices could be better visible for the customers and thus can be easier converted into benefits from SCEM (such as improved image). Customer-side SCEM practices are primarily influenced by public authorities' requirements (e.g. in terms of waste reduction), followed by consumer demand and social awareness and requirements of other parties such as certifying organisations. Supplier-side SCEM is mostly driven by third-party requirements such as requirements of certification systems (like Eco), branch organisations and quality improvement goals. Although the requirements of certification systems might not prescribe supply chain cooperation, certified processors might be more likely to extend their environmental responsibility towards the supply chain. When processors already cooperate with farmers with quality improvement goals, environmental concerns can be more easily incorporated into such a cooperation. Also the influence of public authorities' requirements has a significant impact on supply-side SCEM at $P<0.10$ level.

Table 6.6 indicates that especially supplier-side SCEM practices in supply chains are associated with cost efficiency advantages for meat processors. Customer-side SCEM practices are also positively associated with cost efficiency advantages but the correlation coefficient is smaller and less significant. It is quite interesting that we found no positive association between both types of SCEM practices and market advantage. Also market advantage is negatively related to cost efficiency advantage according to meat processors. This could be due to specifics of the meat industry that trades meat mostly as a commodity. It might be difficult for meat processors to convert their environmental practices into benefits

Table 6.5. Factors that influence supply chain environmental management (SCEM) implementation.

		1	2	3	4	5
1	SCEM: supplier-side practices (environmental cooperation with suppliers, setting environmental requirements for suppliers and being informed about their environmental management)	x				
2	SCEM: customer-side practices (environmental cooperation with customers, working on solutions to optimise transportation and reduce/reuse/recycle the packaging)	0.094	x			
3	Demand and social awareness (consumer demand, social awareness, desire to reduce the cost, improve competitiveness and the firm image)	0.203	0.387**	x		
4	Public authorities' requirements (current and anticipated environmental regulations, financial support by the government, such as subsidies, etc.)	0.245+	0.592**	0.268+	x	
5	Requirements of other parties (certification requirements, requirements of branch organisations, desire to improve quality)	0.510**	0.281*	-0.104	0.310*	x

** = $P<0.01$; * = $P<0.05$, + = $P<0.10$.

Table 6.6. The contribution of supply chain environmental management (SCEM) to business performance.

		1	2	3	4
1	SCEM: supplier-side practices (environmental cooperation with suppliers, setting environmental requirements for suppliers and being informed about their environmental management)	x			
2	SCEM: customer-side practices (environmental cooperation with customers, working on solutions to optimise transportation and reduce/reuse/recycle the packaging)	0.094	x		
3	Market advantage (improved quality, improved firm image, charging higher product price, increasing market share)	-0.156	-0.142	x	
4	Cost efficiency advantage (including lower energy, water, waste and waste treatment cost)	0.289*	0.256+	-0.534**	x

** = P<0.01; * = P<0.05, + = P<0.10.

for consumers. Also consumers do not reward sufficiently more environmentally friendly meat with a market advantage like a price premium or with a larger market share. The latter is consistent with the literature indicating the niche character of sustainable premium-price food markets (e.g. Van Doorn and Verhoef, 2011). The fact that both upstream and downstream SCEM practices are not reflected in image improvements might be explained by the fact that consumers or customers buying meat do not know about SCEM practices in the chain. Such information might be costly to obtain or even not available (De Marchi, 2012). Especially large meat processors place information about their SCEM practices on their website, but customers or consumers might not access this information. While in other industries SCEM practices might result in products with environmental benefits directly observable by customers (e.g. reduced energy consumption like in case of Zhu *et al.* (2008a)), this is hardly possible in the food industry and especially in meat chains due to product specifics.

Conclusion

SCEM practices both on the supplier and on the customer side have a promising potential for cost savings as is illustrated by meat supply chains. Although SCEM practices were not shown to bring market advantage, this could be due to the specifics of the meat supply chains and studies in other sectors or industries could find a positive association between the concepts. In any case, SCEM practices are advantageous for business performance and should be further stimulated: on the customer side by public authorities in combination with consumer demand and third parties, on the supplier side by third parties.

Chapter 7. Risk-oriented animal health and food safety management

Secure animal health is of utmost importance for human health, animal welfare, resource efficiency and the effectiveness of livestock production. Food safety hazards, zoonotic diseases and the use of pharmaceuticals such as antibiotics have potential impacts on human health and wellbeing. Therefore, on-farm health and risk management is crucial to safeguard animal products as well as trade with livestock. Appropriate infrastructures for preventive health management to improve animal and human health and wellbeing are at the core of a 'one health' approach with provides a challenge at all levels of science and business.

This chapter will inform you about:

- which structures have been established and which innovations have been made in risk-oriented animal health and food safety management;
- how animal health management services and health status certification can be organised according to the AMOR principles;
- how and why on-farm biosecurity requirements have changed and how management support should be organised to foster a future one-health strategy.

Keywords: inspection strategies, serological monitoring, hygiene practices, farmer compliance, health-status certification

Summary

The control of diseases is of utmost importance to assure animal and human health and to safeguard the trade of animals and animal products. From the societal and the economic view, it is essential to improve safety aspects of foodstuffs provided throughout the agri-food supply chains that link animals to people. Tools and practical guidelines to ensure supplies of food and feed that are microbiologically and toxicologically safe must therefore be developed and optimised to reduce risks of food borne pathogens, residues and contaminants. This chapter presents risk-based innovations regarding food safety management at slaughter level as well as monitoring and farm auditing schemes provided as a service by Netchain coordinators and organised according to the AMOR principles

DOI 10.3920/978-90-8686-789-9_7, © Wageningen Academic Publishers 2014

(Alliances for the Mutual Organisation of Risk-oriented inspection strategies) on farm level. Efficient control of microbial and chemical hazards in dairy and meat production is the goal of the one-health approach assuring and promoting food security. The starting point is an on-farm biosecurity and Health Management strategy at national, transnational and local level co-ordinated by operating networks and novel health service organisations.

7.1 Risk-based innovations in food safety management

L. Heres[1], A. Hiller[1], G. Schulze Althoff[1] and B. Urlings[2]
[1]VION GmbH, Franz-Rennefeld-Weg, 5, 40472 Düsseldorf, Germany
[2]Wageningen University, Management Studies group, Hollandseweg 1, 6706 KN Wageningen, the Netherlands
lourens.heres@vionfood.com

Food of animal origin requires a special responsibility of everyone involved in the production chain. All market partners must work with great energy to produce safe food and to safeguard the social acceptance for the production of food of animal origin. This effort will be successful when open and transparent answers are given to the legitimate questions of consumers and non-governmental organisations, and when constructive solutions are developed to meet new requirements. Innovations are needed to meet the requirements posed by food safety, animal health and animal welfare issues as well as the environmental impacts of animal husbandry in a constructive way. Accordingly, industry cannot simply continue with its current routine, but has to make efforts to be innovative. Results from research in cooperation with partners from universities are enablers for innovation. These could encompass, for example, innovations that make analyses cheaper, faster and more sensitive, or new methods of data acquisition and data analyses. Besides these technological innovations, there is also a need for organisational innovations in the supply chain towards chain integration and optimisation, and from governmental to private control.

Residues and contaminations in food are regarded as a latent health threat by a considerable number of consumers. The pathways can either be short or direct or longer and include more or less all food chains (Heeschen and Blüthgen, 1993). Figure 7.1 shows the pathways for residues and contaminants within the food chain. The transfer of residues and contaminants between the links 'feed' and 'animal' is characterised by the phenomenon of a so-called carry over. This carry over describes the possibility and extension of the migration of substances from feed into the food-producing animal or a derived food (e.g. milk). The carry-over rate is the percentage of the ingested dose of the chemical to be found in meat or milk (Heeschen and Blüthgen, 1993).

Currently, risk-based or risk-oriented approaches are favoured for different reasons. The benefits of risk-based food control are that the focus will be on aspects of the food chain or processes that pose the highest risk, i.e. the probability of detection will be increased. Based on this approach, control will shift from reactive approaches, i.e. when things have gone wrong, to preventive measures. Risk-based approaches optimise the efficiency of food control and inspection costs which is important since resources are always limited (FAO, 2008).

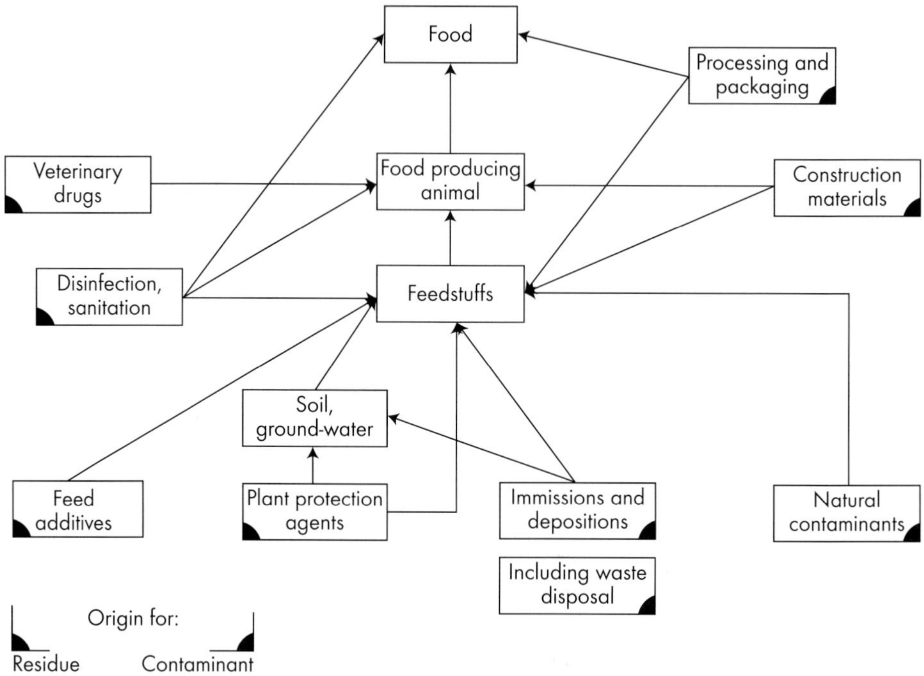

Figure 7.1. Origin of residues and contaminants in food of animal origin. Reproduced from Heeschen and Blüthgen (1993).

Development of food safety control

Procedures that are currently implemented for the control of animal disease and product quality in animal production are also based on science, technology and culture. However, they reflect the status from some decades ago and before specific regulations were implemented. The current meat inspection and milk control legislation are the most obvious examples. But also animal health legislation, rendering practices, and the legislation concerning the veterinary profession were mainly developed during the first 60 years of the 20th century. In pork production, the testing for the presence of *Trichinella* larvae, via trichinellascopy or via artificial digestion methods where each individual carcass is tested, was implemented during this period. At that time, these methods were appropriate techniques adapted to the risk encountered. Another procedure implemented on a large scale was the inspection of each individual carcass (and the offal) during slaughter. In those days, the government was mainly held responsible for food safety and animal health.

Until 1990, food control of one batch of milk or an individual carcass normally only reflected the status of that product but had no effect on the farm or other suppliers in the chain. As from 1980, the systems gradually changed from end-point testing to system and supply chain control. These new systems were mainly oriented towards implementation of high levels of biosecurity at all production stages of animal production.

Two major crises that provoked drastic changes in the food safety policy toward risk-based systems were the Hamburger disease[4] (*Escherichia coli* O157) starting in the USA and BSE starting in the United Kingdom. Policy reports in the EU and the USA released in 2000 concluded that the control of food safety had to be reorganised. Triggered by these incidents, another significant aspect gained in importance, i.e. the responsibility for food safety. This responsibility would no longer reside with the authorities but would by all means be the responsibility of the producer. From a pure legal point of view, food safety has always been the responsibility of the producer but most of the food safety control procedures were developed and carried out by the government, and thus the official authorities were in fact responsible for food safety. These policy documents issued by the EU (White Paper on Food Safety) and the USA paved the way for the new era of food safety control.

In the White Paper on Food Safety (EC, 2000), the basic principle for the food safety policy is mentioned. Here, three principles are of fundamental relevance in the context of designing a new quality management in a company:

1. The prime responsibility of the manufacturer or the production chain for food safety.
2. The control of food safety should be based on science, and include at least risk assessment.
3. Risk analysis and HACCP are important starting points for food safety systems.

Taking into account these basic principles, European legislation has been set up. The basic EU regulation on food, the General Food Law (EU, 2002), the Hygiene Regulations (EU 2004b) and the Regulation on official controls (EU, 2004c) and all related regulations, which are ideally suited to provide the personal responsibility of a Food Business Operator, form the legal framework. For more details, we refer to Chapter 3. These EU regulations and hygiene regulations are aimed at and addressed to companies. Regulation (EC) No. 854/2004 also contains some design rules regarding implementation of veterinary inspections under official supervision, otherwise this regulation is also aimed at companies (EU, 2004c).

It is worthwhile to consider that together with the EU regulations, microbiological criteria were established as well (EU, 2005). They are divided into process and food safety criteria. When exceeding the process criteria, it is necessary to re-evaluate the whole process, including suppliers. Measures must be taken to ensure that the process criteria are met

[4] Infection with *Escherichia coli* O157:H7 that leads to bloody diarrhoea and occasionally to haemolytic uremic syndrome (HUS). Outbreaks have been associated with the consumption of undercooked, contaminated ground beef (Hamburger patties).

in future. When the process criteria are consistently exceeded, the supervisory authority must shut down the facility. Food safety criteria are always directly related to the presence of a potential hazard in the food in question. When exceeding the food safety criteria, the product in question must be recalled. This involves, for example, contamination of end products with *Salmonella*, *Listeria* or histamine. With the introduction of microbiological criteria, more objectivity is introduced as well. Control is not solely based on visual observations and checks, but the performance is measured and trends as to whether a company is stable, improving or having a decreasing performance, can be shown easily. These data can perfectly be used in quality control systems as Key Performance Indicators (KPI s) and in benchmarks between production sites.

The changed role of official verification

The 'new' EU framework and distribution of responsibilities between companies and authorities should be properly understood to establish new science-based food safety control systems. By clearly defining the responsibilities of companies and supervising authorities, the character of the work done by officials changes. The new monitoring framework requires much more knowledge from officials to base the verification programme on HACCP and epidemiology.

These control activities undertaken by the competent authority consist primarily of the following (EU, 2004a):

- Analysis of the functioning of the companies' quality systems and their performance.
- Inspection, especially of production areas, buildings, raw materials, semi-finished products, packaging materials, labelling, cleaning and disinfection.
- Verification of hygiene in the operation.
- Evaluation whether the methods, such as HACCP, comply with EU guidelines.
- Evaluation of the operation's documentation to establish compliance with legislation.

Verification by the competent authority can tie in with the company's own quality systems and their data to assess compliance and performance of the system. This official verification should be able to demonstrate whether the food business complies with its responsibilities. When producers assume their responsibility and can transparently prove their performance and compliance with procedures to the authorities, the authority can step back. Whereas, on the other hand, the authorities should be strict when performance and compliance are poor, or even worse, when companies are not obeying the law.

Due to the new orientation of systems in enterprises and supervision by the competent authority, new possibilities to improve food safety arise. These systems can have different advantages:

- For companies that perform well in the field of food safety, the EU has offered the opportunity to significantly reduce the frequency of monitoring. This accounts for a small part of the investment made in own quality systems.
- Businesses are given the opportunity to design alternative methods for the control of food safety. They can develop, for example, an alternative system for the control of *Trichinella spiralis* on the basis of epidemiological knowledge and control measures in the chain. With the focus now being on the real risks of trichinosis, a higher degree of control can be achieved.
- Combining food safety data from the entire chain can lead to an adaptation of official meat inspection, for example for fattening pigs. In this way, the various food safety hazards can be targeted controls, for example, already at the level of farms, so that the cutting of lymph nodes will no longer be necessary for each animal. This reduces the risk of cross-contamination and food safety is substantially improved.

Examples of risk-based approaches

There is a continuous need for science to design and reassess not only control systems but also the monitoring of these systems. One of the advantages of industries developing new and modernised food safety systems is faster implementation of changes compared to authorities making generalizable approaches for all companies and having to seek agreement before implementation.

New techniques will increase food safety. An example for this is the development of risk-based meat inspection systems (Hiller *et al.*, 2012) where serology is used to identify risk herds (Hiller *et al.*, 2013). This development was enabled by the development of ELISA for the detection of *Mycobacterium avium* (Wisselink *et al.*, 2010). With serological monitoring of every herd supplied, a targeted approach towards these farms leads to preventive measures to reduce the presence of *Mycobacterium avium*, and the abolishment of incision of lymph nodes during meat inspection, causing cross contamination (Hiller *et al.*, 2012), can be scientifically justified.

The same approach can be used for the risk management of toxoplasmosis. Toxoplasmosis is a zoonotic disease caused by the protozoan *Toxoplasma gondii* with serious consequences for pregnant women and pre-natal infants which is also attributable to meat from pig and cattle (Opsteegh *et al.*, 2011). By screening herds on toxoplasma, those herds where a lack of biosecurity results in introduction of toxoplasma, via rodents, cats or feed, can be identified and preventive intervention methods can be introduced. This innovation provides quite a challenge for supervising authorities. According to EU regulations, nothing has to be done in connection with toxoplasmosis so far. When veterinary authorities request from food producers to sanitise carcasses by freezing the meat, they would, however, destroy with this the new pro-active risk-based approach aimed at improving biosecurity at farm level.

Experience was also gathered with a risk-based surveillance approach for residues of antibiotics in slaughter animals. Legislation prescribes random sampling of a population. A new procedure for sampling was designed, based on recent pathological findings in a herd. These pathological observations are a sign of recent infections in the herd and can signal the possible use of antibiotics. When this new procedure for sampling is applied within the same population of slaughter pigs, based on risks, the results show that the chances for finding a positive result increase without affecting the true prevalence of the whole population (Hiller *et al.*, 2012).

Risk assessment showed that the slaughterhouses are critical for the control of *Salmonella*. Owing to this, it is logical that control of faecal contamination is treated as Critical Control Point (CCP) in pig slaughterhouses. Additionally, it was proven that contamination of slaughter equipment can seriously contribute to carcass contamination (Van Hoek *et al.*, 2012; Smid *et al.*, 2012).

In the 90's in Denmark, a *Salmonella* control plan for pigs was set up (Mousing *et al.*, 1997). Afterwards, this plan was copied by other countries, although criteria for risk categorisation and cut-off values differed slightly. Based on the prevalence of positive serological samples, pig herds are categorised in risk categories. The herds in the highest risk category are required to improve. The effect of the screening on herd prevalence in the screened populations seems to be limited. Moreover, Baptista *et al.* (2010) demonstrated that the number of *Salmonella* sero-positive pigs reaching over 200 has no influence on the *Salmonella* prevalence on carcasses. It has also been shown repeatedly that logistic slaughtering of high-risk herds does not reduce carcass contamination (Swanenburg *et al.*, 2001). Not only the effectiveness of screening to reduce carcass contamination is questioned, there are also concerns relating to the methods used. In experimental circumstances, serological screening for *Salmonella* can be based on meat juice. It was found out that in practice, however, the outcome of meat juice is far more heterogeneous than serum collected from blood (Hiller *et al.*, unpublished data).

For Livestock Associated MRSA (LA-MRSA), recent studies have confirmed that LA-MRSA is mainly an occupational risk and not a food safety risk for which hygiene procedures during meat production should be changed (Gilbert *et al.*, 2012). Such findings show that food producers should comply with the general hygienic principles of pig slaughter, but additional measures to control MRSA are not justified when the purpose is to make food safer. A good reason to reduce the infectious load of LA-MRSA at herd level is to reduce the exposure of farmers and other people having contact with live animals.

During a recent contamination incident with dioxin, Heres *et al.* (2010) found out that contamination was detected first by private monitoring systems. Results also indicated that an early warning system can be integrated in the private quality systems of processors of animal by-products, fat and gelatine production. A cost-benefit analysis shows that this is more efficient than sampling carcasses in slaughterhouses (Lascano Alcoser *et al.*, unpublished data).

In conclusion, it is obvious that science has provided the food producers already for years with techniques and knowledge. In view of the changing environment and assigning more responsibility to the producers within the whole supply chain, cooperation between food producers and science is faced with the challenge to make progress towards new systems. Food producers are aware of their challenges and are in need of science concerning food safety that enables an effective and efficient control of food safety. This will be favourable for consumer confidence and provide the industry with instruments to produce wholesome food. Accordingly, industry's sustainability is assured.

7.2 Chain-oriented animal health management services

J. O'Hagan[1], S. Düsseldorf[1], T. Klauke[1], A. Czekala[2] and V. Schütz[3]
[1]University of Bonn, Katzenburgweg 7-9, 53115 Bonn, Germany
[2]Wageningen University, Management Studies group, Hollandseweg 1, 6706 KN Wageningen, the Netherlands
[3]DRV, Deutscher Raiffeisenverband e.V., Pariser Platz 3, 10117 Berlin, Germany
j.ohagan@uni-bonn.de

Animal health management corresponds to on-farm planning, implementation and surveillance of organisational and production-related measures aiming at good animal health. This includes, amongst others, the functioning of technical equipment (drinking troughs, ventilation, heating), stable climate, hygiene, cleaning and disinfection, stable and housing conditions, medication, vaccination regimes, care of the animals, feed (amount, composition, quality), inspection of the animals (behaviour, appearance; performance parameters like weight gain, gestations), monitoring of pathogens, documentation. This shows that animal health management is a multifaceted issue.

An important role in animal health management is assumed by monitoring activities and farm auditing to assess the animal health status. Monitoring is an act of conducting a planned sequence of observations or measurements of certain control parameters to assess whether a defined point in the production process is under control or functioning correctly or shows conformity with market or society demands. The demand for chain-oriented monitoring and auditing systems has increased, and during the past few years many of these systems have emerged on the market, especially in pig production in Germany. This is due to the large differences in the production structure and limited integration of the pig production chain resulting in varying pig health from farm to farm. The main driving factors for the development of the schemes are laws concerning food safety and the responsibility of the food producer, trade and marketing advantages, economic efficiency and a growing public interest. The end customer increasingly demands good animal health and welfare (Moynagh, 2000). Furthermore, there is a high interest of farmers to assess their pig herd health status by participating in monitoring and auditing programmes (Schütz, 2009; Ellebrecht, 2012), especially as a means to increase transparency regarding piglet health for the customer (Nathues *et al.*, 2011). The health status of piglets is an important purchasing criterion for fatteners (Petersen *et al.*, 2010).

The programmes can be provided as services by service providers and can be organised according to the AMOR principles (O'Hagan *et al.*, 2013). The schemes represent a complex coordination service consisting of local farm audits and monitoring activities comprising laboratory investigations on pathogens, and, based on the results, status classification and communication (Petersen *et al.*, 2010). The participants mostly receive a health status certificate.

Currently, monitoring and auditing services are provided by various service organisations, such as Netchain coordinators, that help farmers with this complex task.

Function of the Netchain coordinator

Netchain coordinators in food supply chains are 'organisations with bundling functions and offering services to reduce interfaces for the farmers by coordinating and supporting product and information flows' (O'Hagan *et al.*, 2013).

For animal health monitoring and auditing schemes, repeatable activities have to be performed. Coordinators monitor various disease parameters on-farm as well as the level of implementation of measures concerning animal health, assist the farmers in improvements and facilitate realisation of defined optimisation protocols. The following service activities are frequently offered in combination by Netchain coordinators to service users to achieve good animal health (O'Hagan *et al.*, 2013):

- organisation of an intercompany information exchange;
- organisation and realisation of monitoring activities;
- organisation and realisation of audits and consultations;
- support with data processing;
- issuing of health status and company-compliant certificates.

Netchain coordinators may be partners from within the Netchain as well as external institutions which professionally offer coordination services (Wohlgemut, 2002). Enterprises acting as Netchain coordinators require contextual knowledge and a good organisational overview of the mechanisms of the existing network (O'Hagan *et al.*, 2013).

Potential Netchain coordinators in the pork supply chain include private organisations such as farmers' cooperatives, producers' associations, livestock traders, livestock marketing corporations, slaughterhouses, meat companies, retailers, and public institutions such as animal health services (Ellebrecht, 2008; Schütz, 2009; Brinkmann and Petersen, 2010).

Amongst these potential Netchain coordinators, numerous service providers offer health monitoring and auditing services, such as the following:

- *Public institutions*. Pig Health Service Bavaria, Pig Health Service of Lower Saxony and the Danish Meat Board
- *Private organisations*. PigMatch from the Netherlands, the German slaughter company Westfleisch and the Animal Health Agency TiGA in Germany.

AMOR principles

Chain-oriented animal health monitoring services documented with a certificate can be successfully organised in a mutual approach and risk-oriented manner between Netchain coordinator, piglet producer and pig fattener according to the AMOR principles (details, see O'Hagan *et al.*, 2013). This acronym stands for 'Alliances for the Mutual Organisation of Risk oriented inspection strategies' and has already been introduced in Chapter 4. AMOR is a novel concept to consolidate resources and create benefits for each participant while reducing efforts and costs through jointly performing inspections. Regarding animal health services, the AMOR concept proposes a centrally organised method of company auditing and monitoring for certification of farm animal health. Valuable information is created and available at the piglet producing and fattening stage as well as in the Netchain coordinator company. The inspection strategy resulting in a health status certificate adheres to the following general procedures. The piglet producer registers for the programme with a Netchain coordinator. The Netchain coordinator then commissions a veterinarian to conduct animal testing and sampling. The veterinarian collects all relevant samples and sends them to an accredited laboratory for standard analysis according to (inter)nationally recognised methods. The veterinarian conducts an audit to assess the general conditions of the farm and to determine the state of animal health and animal health management. With the aid of the laboratory test results, the Netchain coordinator issues a health status certificate to the piglet producer. The producer can take appropriate measures according to these results and furnish the health certificate to the fatteners and/or other interested stakeholders. Figure 7.2 summarises the characteristic aspects of an inspection strategy according to the AMOR principles including responsibilities and tasks, information and communication structures, and benefits and costs for the individual parties. The principles can easily be applied to other agri-food supply chains.

Characteristics and examples of chain-oriented health management systems

The health status certification process includes different criteria depending on the certificate-issuing service provider since universally accepted methods and values are lacking. However, analysis of the health status including auditing has to be based on nationally and/or internationally standardised analytical methods, cut-off points or management practices. The service provider will judge a company's performance in meeting predefined criteria and will subsequently assign a health status certificate to that piglet producing company. For certification, systematic sampling for laboratory investigations is a necessary precondition in the producer's company (O'Hagan *et al.*, 2013).

The health status of a farm is important for three purchase decisions (Figure 7.3) when trading animals from one farm to another, from fatting to slaughtering or during processing.

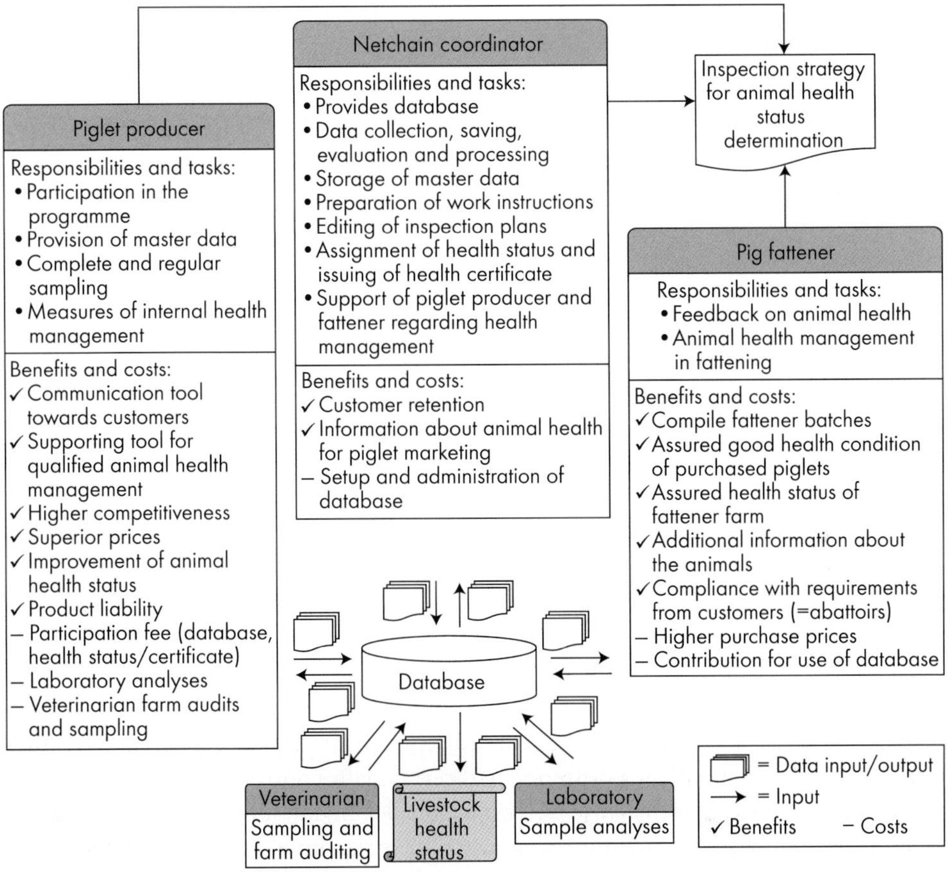

Figure 7.2. Characteristics of AMOR (Alliances for the Mutual Organisation of Risk oriented inspection strategies) inspection 'livestock health status' (O'Hagan *et al.*, 2013).

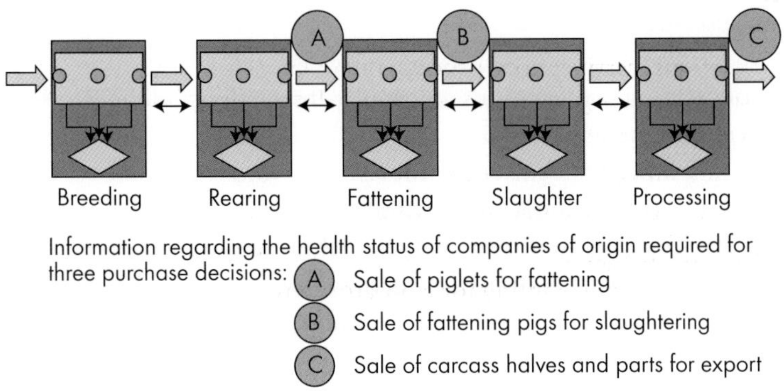

Figure 7.3. Inspection points in supply chains according to three purchase decisions.

An inspection at point A corresponds to the health status certificate described in this chapter. At point B, a monitoring scheme of pigs for slaughter on various disease parameters can be performed to assess animal health and in the long run to tackle specific known health problems and improve animal health. Such a scheme is suggested by Düsseldorf (2013) using a monitoring system based on serological tests of slaughter blood. At point C, meat juice samples can be used to inspect meat samples with biosensors on chip basis and to draw conclusions on animal health, animal welfare and product safety (Klauke, 2012). Subsequently, existing chain-oriented health monitoring and auditing systems in the pork supply chain are presented.

Denmark and the Netherlands each have country-wide central systems for pig health management. The first nation-wide pig health management system was developed in Denmark in 1971 (SPF-system[5]; specific pathogen free). It basically represents the origin of all following developments of health management systems for the pork production. Since its establishment, this system has given a guarantee of success to Danish pork meat exports to the EU and to third countries. The SPF system considers varying needs through different security levels but the key element – strict controls of compliance with regulations and hygiene standards – guarantees the success.

The first system in the Netherlands (Biggen Pas)[6] was founded in 2009 to support information transfer between rearing and fattening farms. Biggen Pas structures which type of information should be transferred without actually controlling use and content of the data. Farmers can download the freely available document, fill in the information and confirm the correctness of the information with their signature. There are no monitoring activities or laboratory tests included and no joint database is provided. Since 2012, farmers from the Netherlands and from Germany can use the PigMatch system which is mainly focused on breeding farms (www.pigmatch.eu). Participants can chose between different levels of participation. The levels vary in frequency of monitoring and inspections. At each level, a specific number of pathogens are tested. The results are presented on the website for interested customers.

In Germany, pig trading organisations are mainly responsible for the development of pig health management systems, like:

- ZNVG Vermarktungsgemeinschaft für Zucht- und Nutztiere e. G. (marketing association for breeding animals and livestock);
- EGF Erzeugergemeinschaft für Qualitätsferkel im Raum Osnabrück e. G. (producer association for quality piglets in Osnabrück); and
- Erzeugergemeinschaft Südostbayern e. G. (producer association of Southeast Bavaria).

[5] http://www.spf-sus.dk.

[6] http://biggenpas.mijnpve.nl.

Another system called Westfalenpass has been established by Westfleisch e. G., a leading slaughterhouse. The aforementioned company-driven systems are restricted to members on farm level and to contract farmers. The systems include a regular monitoring scheme on pathogens. The aim is to simplify trade between piglet producing and fattening farms and between fattening farms and slaughterhouses. The systems are used as an instrument for customer relationship management.

The systems in Germany only have regional relevance. Only TiGA (Tiergesundheitsagentur e.G.; animal health organisation), founded in 2010, offers a system that is not restricted to a certain region or certain trade partners. This system includes specifications for testing parameters, number of samples, laboratory methods, and schemes for auditing and checklists. The following programmes primarily have the goal to improve the health of pigs and include public veterinary services:

- BayPHV (Bayerisches Produktionshygieneverfahren – Bavarian Procedure for hygienic production);
- EVH-Select Screening (Ems-Vechte-Hase-Select Screening);
- The project 'Gesunde Tiere – Gesunde Lebensmittel' (Healthy Animals – Healthy Food); and
- The project 'Veredelungsland Sachsen 2020 – Fitness Monitoring' (pig processing in the federal state of Saxony).

Recently, the Southeast Bavarian monitoring system based on serological tests of blood sampled during slaughtering was developed. Recent findings show that farms with lower serological values in the slaughter blood used less antibiotics, had better growth and better feed efficiency (Düsseldorf, 2013). This indicates that serological screening of slaughter blood can be used to improve pig health and to develop easy-to-use benchmarking tools for pig farmers to assess their animal health status (Düsseldorf, 2013).

For the animal health management systems, the French intercompany chain information system Sanibase can be used which enables secure access to relevant information regarding animal health. Sanibase gathers information in one place and centralises information relating to actual on-farm health status, such as laboratory information, cooperatives information, technical information (weight gain, feed consumption, etc.), prescriptions of health veterinarian communities and visit protocols and/or slaughter data. Authorised users can instantly share the same information (http://www.q-porkchains-industry.org/?id=324).

Harmonisation of systems

Pig health management systems have not been harmonised yet as their diversity stresses. However, first attempts in order to harmonise different pig health management systems to increase comparability have been made (Petersen *et al.*, 2010; Schulte-Wülwer, 2012; Nathues *et al.*, 2011). These attempts mainly concentrate on monitoring details such as the kind of pathogens, methods of sampling and laboratory methods. They have not been put into practice yet.

Harmonisation of pig health management systems, at least to a certain extent, would offer a number of opportunities. These opportunities include supra-regional improvement of the pig health status of herds resulting in the prevention of epidemic diseases as well as the reduction of antibiotics use. Cross-border trade is simplified when market restrictions are reduced if production requirements remain similar. In addition, a harmonised system could reduce costs due to economies of scale and size and also decrease transaction costs since buyers of piglets or pigs can choose from a bigger supply base and suppliers can sell to a larger market. Furthermore, a common database is able to improve the information flow. It would be possible to use the results of (harmonised) monitoring in the sense of a public-private-partnership in co-operation with the veterinary authorities in case of crises affecting animal and public health. Moreover, the existing communication structures can be used to exchange further information. A harmonised system also offers the opportunity to enable access to other databases through interfaces.

The challenge currently being faced is that there are still more data available at farm level that could be combined with pig health data such as performance data, health data about the status of living pigs and after slaughtering. If information from different systems is combined, the farmer would get a better overview about the health status of his herd and thus advisory services can be improved. In order to meet the demands of the different stakeholders and to increase acceptance, chain-oriented/interplant aspects should be considered more intensively.

Improvement of quality methods in animal health services

Ellebrecht (2012) explored in her study how Netchain coordinators of pork supply chains can extend and improve their coordination services regarding intercompany health management by using preventive quality management tools. Figure 7.4 shows how the most important methodical steps of preventive quality management methods, i.e. HACCP, FMEA and Six Sigma, can be meaningfully combined for the core processes in inter-enterprise health management of service providers in the food chain. The concept consists of three parts. After the preparatory phase in which a quality management team involving farmers, veterinarians and advisors is defined, responsibilities are determined. Part 2 contains activities carried out with the support of the team as a whole (process description, objectives, definition of potential risks or deviation, cause-and-effect analysis, definition of control points). Realisation of these methodical steps is necessary when introducing quality management methods and when there are changes in the processes. The final methodical steps are summarised in Part 3. They are carried out in a process relaxed form and thus continuously. This only involves the individually responsible persons (monitoring of compliance with control points, risk assessment, implementation of measures, verification).

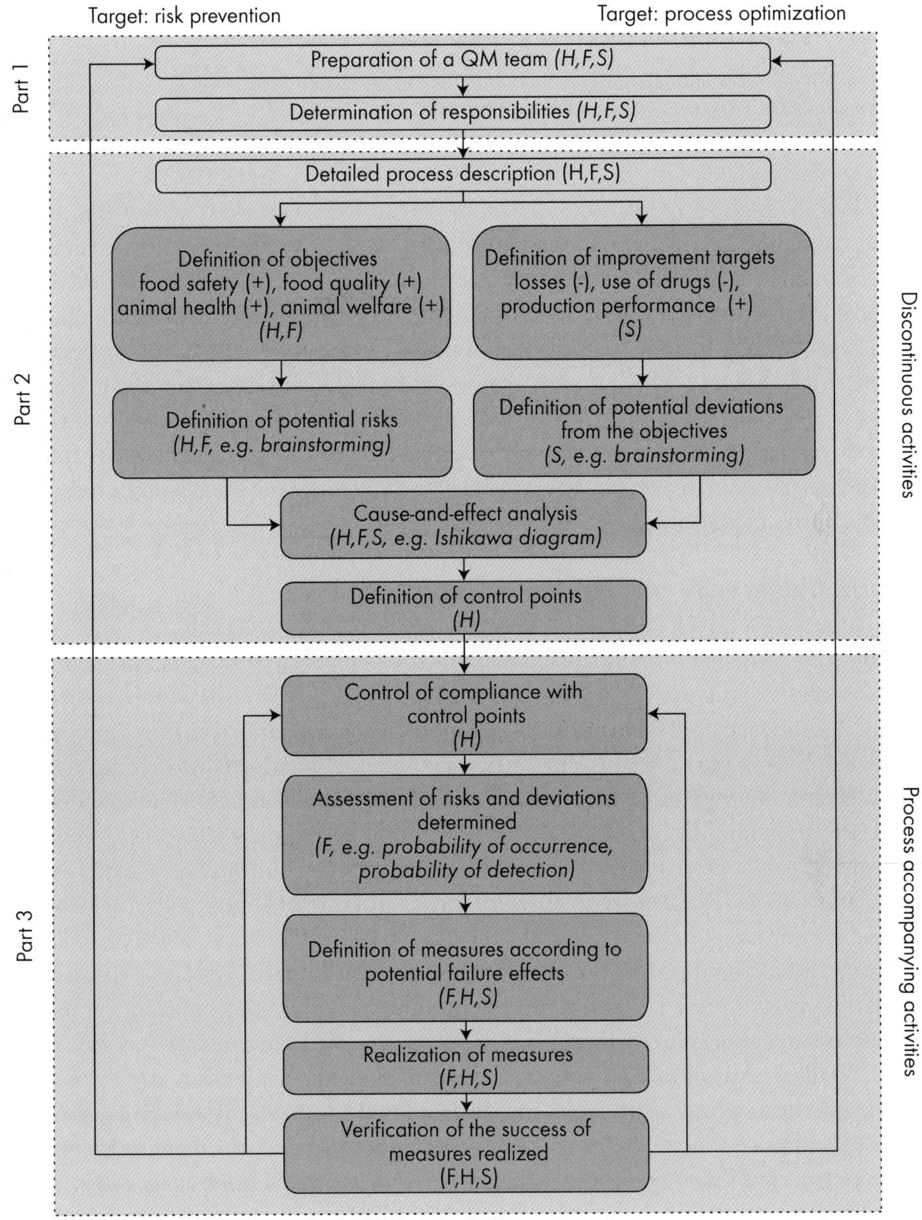

Inidication of origin of introduced methodical steps:
(F) = Failure Mode and Effects Analysis, *(H)* = Hazard Analysis and Critical Control Points, *(S)* = Six Sigma

Figure 7.4. Procedural steps regarding the use of preventive quality management methods in processes of intercompany health management (Ellebrecht, 2012).

7.3 On-farm biosecurity management

T. Wilke, S. Schulze-Geisthövel and B. Petersen
University of Bonn, Katzenburgweg 7-9, 53115 Bonn, Germany
tilman.wilke@uni-bonn.de

It is a basic assumption that farms with a high level of biosecurity contribute less to the spread of contagious animal diseases and zoonotic pathogen load in the food chain. Within the last decade the terms 'Biosecurity' and 'On-farm biosecurity' became increasingly popular. This section deals with on-farm biosecurity as a concept for the reduction of biological risk in intensive animal holdings. The principles of on-farm biosecurity are becoming increasingly important in regions with a high density of animal production. Practical difficulties are described and illustrated with empirical findings from recent surveys. Finally, it is explained how Netchain coordinators can support farmers by implementing the novel concept of on-farm biosecurity management.

On-farm biosecurity and animal hygiene

The term 'Biosecurity' is always associated with the control of biological risks. Different disciplines and leading international organisations like FAO, WHO, OIE have spawned different interpretations of the term Biosecurity[7]. In the field of animal health, Biosecurity is used as a summary term for all measures, rules and recommendations that help to safeguard either a geographic region or a single animal holding from contagious animal diseases. Hirsch (2010) underlines these different viewpoints by dividing Biosecurity measures into 'regional' and 'operational' measures. In addition, some authors distinguish between *external* and *internal* Biosecurity to differentiate whether spread between farms (external) or spread within farms (internal) is tackled (e.g. USDA, 2001; Wells, 2000). In a similar way, the terms Biocontainment and Bioexclusion describe two aspects of Biosecurity, namely the protection of farms against external threats (Bioexclusion) and the internal hygiene measures to prevent further spread of pathogens (Biocontainment) (see FAO/OIE /IBRD, 2010). From a quality management perspective, this distinction is less important than the question: who is (and feels) responsible for Biosecurity? Protecting geographic regions or administrative districts against biological risks is the primary responsibility of public veterinary authorities. Safeguarding an individual animal holding is, above all, the task of the farmer himself. This leads us to a definition of the term 'on-farm biosecurity measures': the term 'on-farm biosecurity measures' covers actions and facilities that fall

[7] FAO has a very broad understanding of Biosecurity as a synonym for biological risk management. WHO uses the term Biosecurity in the context of laboratory safety guidelines for the protection of laboratory workers (Biosafety) and the containment of hazardous pathogens (Biosecurity).

Box 7.1. Terms and definitions.

Biological agent	Organisms, microorganisms and their toxins
Biological hazard	Biological agent with the potential to cause harm to public and animal health
Biological risk	Combination of the probability of occurrence of a biological hazard and the severity of that harm to public and animal health.
Biosecurity	Entirety of all goals, concepts, processes, measures, facilities and political frameworks that help to minimise biological risks.
On-farm biosecurity measures	Entirety of all purposive preventive activities and facilities supporting Biosecurity that lie under the farmer's responsibility.

under the responsibility of the farmer and are intended to shield the farm from biological risks.

On-farm biosecurity measures take effect upon the vertical and horizontal dimension of the food chain (Figure 7.5). On the vertical dimension, farms are mainly protected against biological risks that stem from upstream of the food chain (e.g. buying feedstuff). And farms with a high Biosecurity level directly reduce the transmission of biological risks downstream of the food chain towards the next harvesting or processing stage (first order) and indirectly affect biological risks in further stages (second order). On the horizontal

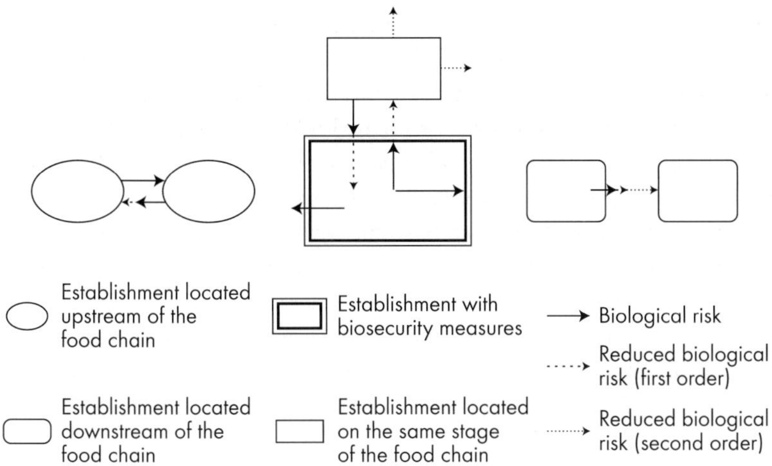

Figure 7.5. Principle of risk reduction along the food chain and within one stage of the food chain through Biosecurity measures.

dimension, farms protecting themselves against being affected as well as protecting other animal holdings with whom they are interrelated (via trade, neighbourhood, road traffic, visitors, veterinarians, colleagues).

Most on-farm biosecurity measures are preventive measures. From an epidemiological point of view, measures are considered preventive if they are able to reduce the effect of a known risk factor or to avoid or to eliminate that risk factor (Wilke *et al.*, 2012a). Referring to the well-known terminology of human medicine, most on-farm biosecurity measures belong to the area of 'primary prevention' as the majority of these measures have the goal to inhibit infections completely. Table 7.1 lists examples of recommended on-farm biosecurity measures for livestock farming.

For some readers the term (On-Farm) Biosecurity might sound new, or they might raise the question how Biosecurity is linked with well-known concepts such as animal hygiene. Animal hygiene covers all influences from the environment to the animal which can affect animal health. Besides biological hazards, this includes non-infectious causes of disease

Table 7.1. Examples of on-farm biosecurity measures for livestock farming.

Controlled purchasing of animals	New animals are purchased from as few suppliers as possible and arrive in as few deliveries as possible. Only animals with a valid animal health certificate are bought.
Isolating pen and quarantine	New animals are placed in a separated pen and stay in quarantine for at least 30 days. Animals in quarantine are examined and tested at mid-term and at the end of the quarantine period.
Clean and unclean zones	The site of the farm is surrounded be a solid fence and a closed gate. The site is structurally and organisationally divided into a clean area (white zone) and a potentially contaminated area (black zone). Well-maintained hygiene locks are the only way to get from one zone to the other.
Controlled visitors	Visitors are registered (date, time, name, place and date of previously visited farm). Visitors enter the site only with protective clothing. Visitors must take a shower and change clothes and shoes before entering a clean zone. All visitors (including vets) use the hygiene locks.
Staff hygiene	Farm workers are not allowed to work at other animal holdings or to operate their own animal holdings. All workers change clothes and shoes after they have visited other farms. No short-cuts are used to bypass hygiene locks. Everybody washes their hands before and after entering a stable. Boots are scrubbed before they are put into disinfectant solution. The disinfectant solution is replaced every day.
Safe vehicles	A loading platform at the border of the clean zone is used to load animals and is cleaned and disinfected after use. Routes of own and foreign cars are located and non-crossing routes are determined. Only clean and empty transporters are allowed to enter the site.
Deliveries at the border of the farm	Vehicles that probably visited other farms before (e.g. feed suppliers, carcass collection service) do not enter the farm but deliver or collect their goods at the border of the site.
Protection from wild animals	Fences are intact and periodically checked. Rodent who can act as disease vectors are under control. Birds cannot enter the stables or feed storages.

like heat, light, noise, climate and technical infrastructure (see Hoy *et al.*, 2006). On-farm biosecurity and animal hygiene are overlapping concepts and refer to each other: without considering animal hygiene, the goals of on-farm biosecurity cannot be reached. And without the measures and rules of on-farm biosecurity, it is impossible to maintain good animal hygiene in intensive livestock production.

On-farm biosecurity measures become increasingly important because farmers in countries of the European Union are confronted with new requirements from private food standards, veterinary authorities, slaughter companies, animal welfare programmes and other societal stakeholders to comply with strict on-farm biosecurity. On-farm biosecurity is used as a criterion in risk-based surveillance and inspection systems like cross-compliance checks, risk-based meat inspection, veterinary inspections and third-party audits of voluntary private food standards. For example, as a result of a recent research project called SAFEGUARD, Biosecurity assessment scores from regular audits carried out by a private quality assurance system are used by some district veterinary offices to substantiate their decision regarding which farm is to be inspected next (GIQS e.V., 2012). Usually, on-farm biosecurity measures are applied in order to protect a farm from naturally occurring risks. During the past decade, however, public and private policy makers of the agriculture and food sector realized that the possibility of man-made threats to food safety, animal health and the environment (e.g. deliberate contamination of food) must not be neglected. On-farm biosecurity measures are the first frontier within the scope of the agricultural and food sector were the chances of such an attack being successful (and therefore the overall vulnerability of the remaining food chain) can be reduced. Hence, on-farm biosecurity is a fundamental part of any food defence strategy.

Practical difficulties and empirical findings

On-farm biosecurity is associated with a high level of uncertainty (Gramig, 2008) for those setting the requirements (public authorities, private quality assurance systems) and for those being responsible to implement measures on their farms. Both groups face the dilemma that on-farm biosecurity principles sound reasonable and logical in theory but include a number of practical problems and hurdles. The main chances and difficulties that come with on-farm biosecurity can be summarised as follows:

Chances:
- Protection from highly contagious animal diseases and therefore protection from direct losses and indirect economic effects of epizootic outbreaks and animal welfare issues during an epizootic outbreak.
- Protection from endemic production diseases and therefore reducing morbidity and supporting high biological performance, animal welfare and reduced medical treatment with less antimicrobial resistance and residues.

- Protection from pathogens that are part of animal health certificates and therefore retaining the possibility to serve customer needs.
- Reduction of the occurrence of pathogens in the remaining food chain and therefore reduction of the possibility of liability claims.
- Support of community-protective effects if all farms of a region or a supply chain maintain high Biosecurity levels.
- Protection from the loss of valuable individual animals (e.g. boars, bulls, cows).

Difficulties:

- The outcome of measures is not reliable, is highly uncertain and will only become apparent in the long term.
- Implementation costs do not directly affect a farm's operating result and the effect of measures is not directly observable by the farmer.
- The measures lead to additional daily work and proper application might disrupt existing working routines.
- The measures might require considerable inventory investments and might require fundamental changes in the production system.
- Failures in execution of measures are easy to overlook or to disregard, and proper application requires to train and advice workers continuously.
- Some measures require coordination and agreement with other parties (vets, employees, suppliers, customers) and might result in additional documentation and paperwork.
- The measures might support the protection of other farms (e.g. neighbours, trade partners) but the own farm may still suffer from pathogen intrusion.

It is obvious that most on-farm biosecurity measures come off badly, when the (quite concrete) difficulties are compared to the (quite vague) chances. As a consequence, On Farm-Biosecurity measures are mostly well-known but are too often not applied consequently by famers. It must be kept in mind that most measures consist of a structural and an organisational part differing greatly regarding risk reduction, costs, responsibility, verifiability, reliability (e.g. hygiene locks consist of physical equipment combined with compliance with standard operating procedures; access controls consist of gates combined with the strictness of security personnel). All measures contain a technological part (e.g. well-maintained disinfection equipment) as well as a managerial part (e.g. workers are regularly instructed in proper disinfection).

Empirical findings from a recent survey among German pig farmers illustrate the deficiencies concerning on-farm biosecurity[8]. Half of the farmers 'always' applied only 6 or less out of 13 on-farm biosecurity measures. 23 out of 53 farmers stated that they do not maintain

[8] Results stem from a web-based survey with two samplings in 2012 and 2013. Unless otherwise stated, the figures relate to a total number of valid responses of n=53.

a hygiene plan for their farm. Although other farmers mentioned that they have up to 20 vehicles (min: 1.0, max: 20.0, median: 6.5) per month on their premises, 28% of the farmers had no idea how much cars and truck enter their farm. Concerning visitors, the situation was better since 91% of the farmers knew how many visitors enter their stables per month (min: 1.0, max: 10.0, median: 3.0). Figure 7.6 shows the distribution of answers to the question how consequently 13 different on-farm biosecurity measures are applied on the respondent's farm. The large amount of 'not possible' and 'never' answers reflects a big potential for advisory and management support.

Not all parties involved in professional livestock production have the same attitude towards preventive concepts like on-farm biosecurity. But in the end, it is a matter of choice of measures and the farmer's compliance that determines a farm's on-farm biosecurity level and as a consequence the Biosecurity of a region or a production chain. The commitment

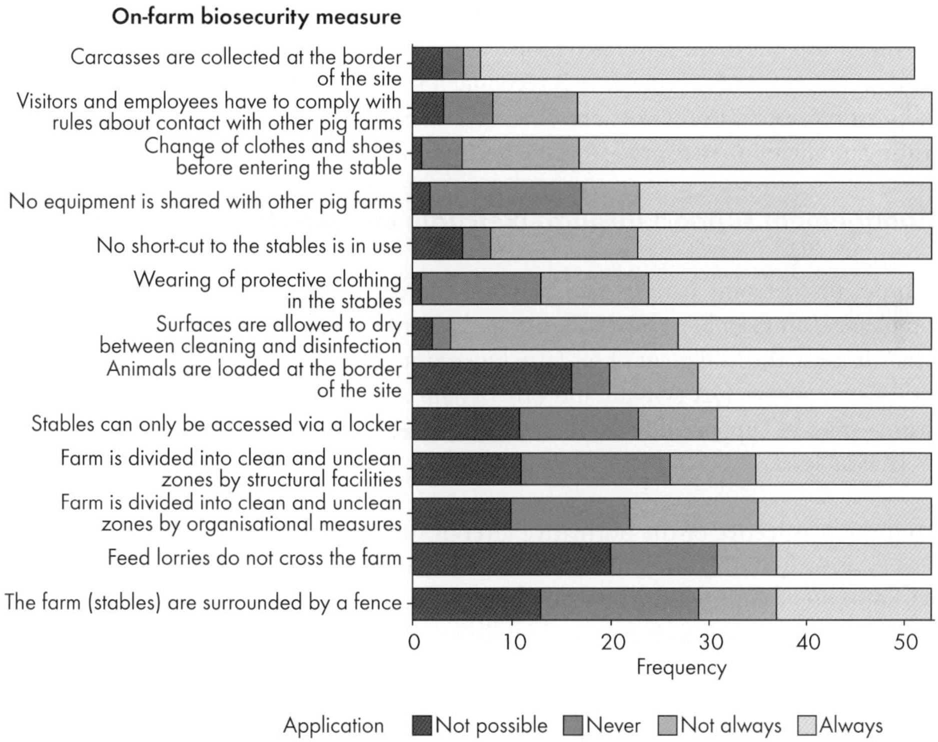

Figure 7.6. Compliance with on-farm biosecurity measures.

Stacked bar chart showing the frequency of answers of a sampling including 53 pig farmers to the question how consequently 13 different on-farm biosecurity measures are currently applied on their farm. The measures are ordered by frequency of the answer 'always'.

to fully comply with on-farm biosecurity recommendations depends on a number of factors. Based on the assumption that perceived practice characteristics are the main driver for farmer's behaviour to adopt and implement measures (Reimer *et al.*, 2012), assessing on-farm biosecurity measures from a farmer's point of view will help to understand the difficulties regarding implementation and to create quality management concepts for continuous improvement in on-farm biosecurity.

In order to gain insight into the preferences and possible barriers for compliance, a survey among German pig farmers was conducted. Farmers were asked to rate a set of 23 different properties of on-farm biosecurity measures according to the importance of these properties for the success of their business (using a 7-point Likert scale from 1-irrelevant to 7-crucial). 18 out of 23 properties had a median rating of important (5) or more. According to the results of that survey, the two most important properties of an on-farm biosecurity measure are that it directly assures the farm's operational result and that the effect of a measure is reliable and predictable for the farmer. Interestingly, the respondents seemed to be well aware of the fact that on-farm biosecurity might require some additional work, as 'additional work load' was one of the least rated properties. But the pig farmers also found it important (median rating ≥5) that a measure is recommended by a trustful resource, that its success and possible application errors are observable for the farmer, that a measure fits to the existing production system and working routines and that legal requirements are fulfilled.

Management support through Netchain coordinators

In order to overcome barriers and practical difficulties regarding implementation and compliance, on-farm biosecurity should no longer be a matter of concern for individual farms. Instead, the fulfilment of on-farm biosecurity requirements – established by public authorities, private QA systems and other chain actors from the normative of the food chain – must be supported at the intermediate level of the food chain by entities working above farm level.

From the perspective of the normative level of the three-level model of the food chain (Petersen *et al.*, 2007), on-farm biosecurity contains three main tasks: first, to formulate risk-appropriate on-farm biosecurity requirements for the farm-level. Second, to exert pressure on the farmers to behave in compliance with the requirements set. And third, to prove which on-farm biosecurity measures are actually applied. The latter can reach from individual farm certificates up to aggregated statistics for a large number of farms. For the public side, task one and task two are addressed by a legislation that traditionally applies to large groups of farms (e.g. regulations that apply to all pig holdings in a country). Regarding task three, veterinary authorities have only very limited capacities to inspect measures undertaken at farm level or to monitor them. On the private side, the main instrument to fulfil this task are voluntary quality assurance systems (QAS) based on self-checks with third-party audits. Regarding task one, it is theoretically possible that QAS develop more individual or stricter on-farm biosecurity requirements than public authorities. But in fact most QAS use public regulations as a basis. Task two and three are combined within the QAS's auditing system. Whether these audit data are meaningful and can be used in an aggregated way depends on the auditing scheme and the information management system of the QAS.

On the operational level, the population of animal holdings is quite diverse regarding the necessary on-farm biosecurity level, and the compliance of farmers depends on individual preferences regarding properties of measures. Therefore, a one-size-fits-all approach in regulating on-farm biosecurity at farm level is doomed to fail, farmers rather need tailor-made concepts according to the risks occurring at their farms and adapted to their production system. But the animal health effects and the economic effects of on-farm biosecurity measures are (still) too indirect and long-term in nature resulting in individual farmers spending a lot of working time selecting and testing measures, investing money in facilities or changing their production system and regularly checking for correct application and possible improvements. Hence, these tasks need to be transferred to an entity that has the required knowledge and resources and is representative of a more homogenous group of farms than law could ever be.

Existing Netchain coordinators[9] are in the right position to fulfil the new tasks of on-farm biosecurity Management (Figure 7.7). In practice, Netchain coordinators can be e.g. existing cooperatives, producer associations or large veterinary practitioners. They need to have in common that they group a number of livestock farms which they will represent towards the normative of the food chain. The linkage between Netchain coordinator at the farms within 'their' network should to be stable and trustful in order to make coordinating tasks possible.

The main tasks of Netchain coordinators within the concept of on-farm biosecurity management (OFBM) are:

- to consolidate on-farm biosecurity requirements of public authorities and private QA systems and other chain partners from the normative of the food chain in order to anticipate and resolve disagreements;
- to translate rather general requirements into practical recommendations for individual farms according to their risk profiles;
- to advise groups of farms or individual farms to choose and implement on-farm biosecurity measures that fit their needs and limitations;
- to assure a common understanding and harmonize implementation of on-farm biosecurity within their 'compartment' (training);

[9] For more details about the role of Netchain coordinators, please refer to Section 7.2 of this chapter.

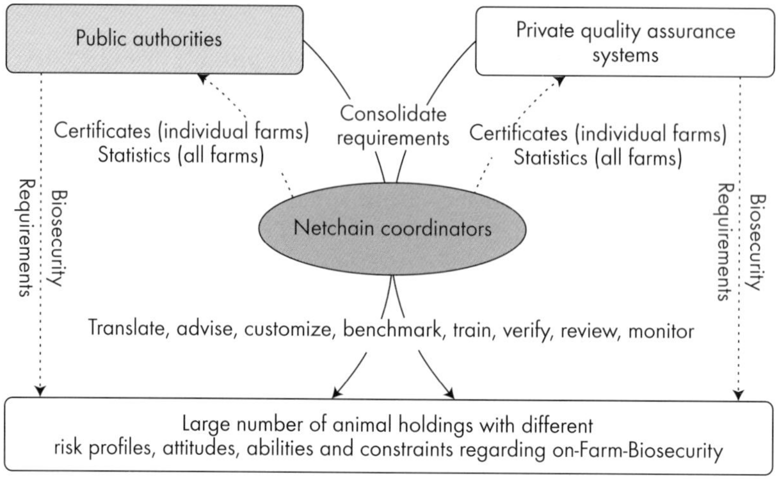

Figure 7.7. Role of Netchain coordinators within the concept of on-farm biosecurity Management (OFBM).

- to verify and compare correct implementation of measures and giving feedback and advice to the farmers (benchmarking);
- to monitor the on-farm biosecurity status of member farms;
- to support regular revision and continuous improvement of a farm's on-farm biosecurity measures; and
- to prove the fulfilment of on-farm biosecurity requirements of farms towards public authorities, private QA systems and other chain partners, either in the form of individual certificates or as aggregated statistics.

Because the success of on-farm biosecurity stands or falls with compliance of the farmer, Netchain coordinators should consider the use of a powerful incentive system. Incentive mechanisms for food safety and animal health have proven to be successful (Van Wagenberg, 2010). Regarding on-farm biosecurity, however, further research is needed in order to understand most of the drivers for adoption of and compliance with recommended measures (Wilke *et al.*, 2012b).

Although the concept of on-farm biosecurity management was developed for the stage of livestock farming, it can be reasonably translated to other food chains as well (e.g. in the fruits and vegetables sector). And the fundamental tasks of Netchain coordinators of the primary stage can (with slight modifications) also be transferred to other stages of the food chain where Netchain coordinators can be identified. Because Biosecurity is relevant on different geographical scales from local to regional to international, Netchain coordinators can also assume an important role in the establishment and the observance of so-called high-risk areas with special requirements regarding Biosecurity. This can be a corner-stone of supra-regional or international Biosecurity strategies which will be necessary to tackle future animal and veterinary public health problems.

Joint risk management tools

A combination of both, i.e. improved on-farm biosecurity measures and monitoring activities to judge the efficacy and the efficiency of measures, would be advantageous from the perspectives of quality and animal health management. Service providers can support farmers during implementation and application of novel prevention approaches. As mentioned in Section 7.2, Ellebrecht (2012) elucidated that service providers need preventive quality management methods to improve their services. According to this, a combination of improved on-farm biosecurity measures and monitoring activities requires joint risk management tools to be used by the farmers. Joint risk management tools of suppliers and customers in the chain have a high potential to be drivers for implementation and persistence of innovations in food production chains. Risk matrices are an example of a valuable tool to illustrate the importance of health management strategies at farm level. They are easy to use, easy to explain and effective in promoting health services (Düsseldorf, 2013). Typically, a risk matrix has two dimensions, the severity and the probability of an adverse event. According to the values of probability and severity, an individual farm is placed in the coordinate system of the risk matrix. For easier interpretation of the location within the risk matrix, defined areas of the coordinate systems might be given meaningful names (e.g. 'high risk group', 'low risk group', etc.). Figure 7.8 shows an example of a risk matrix from a recent research project. It illustrates the risks of a sample of 78 fattening periods of 20 pig farms for respiratory illness (severity) according to a serological indicator from health monitoring (probability).

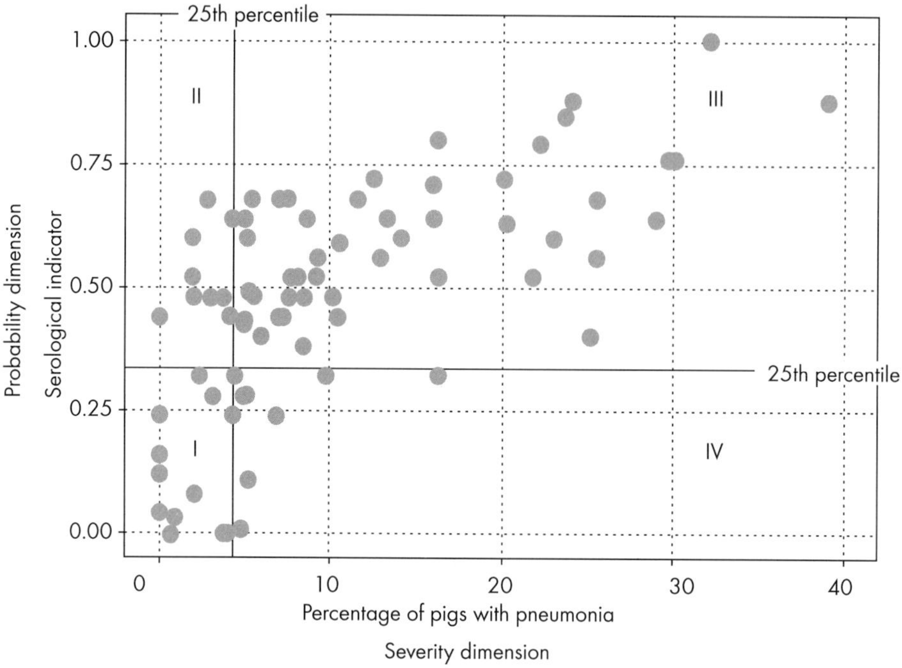

Figure 7.8. Enzootic risk matrix.

Ranking of 78 fattening periods of 20 pig farms according to their enzootic risk potential. The groups are divided by the 25th percentile of each dimension. Group I = 'absence of high enzootic risk', Group II = 'controlled high enzootic risk', Group III = 'uncontrolled high enzootic risk', Group IV = 'less pathogen pressure and higher respiratory problems'. 'Serological Indicator (SI)' represents the average proportion of positive samples (P) in a fattening period in an array of serological tests for antibodies against 5 different pathogens, calculated as SI = (PPRRSV + PPCV2 + PMhyo + PSIV + PAPP2) / 5 / 100. 'Percentage of pigs with pneumonia' is calculated as share of carcasses of each fattening period that show middle-grade or high-grade pneumonia during post-mortem meat inspection. Representation based on Düsseldorf (2013: 46).

Chapter 8. Experiences in intercompany quality management

Introduction of a QM system can take various forms. Above all, in the chain of food-producing companies it is necessary to assign intercompany tasks in quality and risk management and to coordinate them. In the following, sector-specific organisational structures and QM methods will be explained.

This chapter will inform you about:

- which organisational structures assume coordinating tasks;
- how suppliers in the food industry are assessed;
- how systematic risk management can be designed.

Keywords: intercompany quality management, sustainability, organisational structures, coordination services, supplier assessment, preventive risk management

Summary

Customer-supplier relationships in the agri-food industry are very complex and heterogeneous. Therefore, the possibilities for designing intercompany QM systems are just as diverse. The most important quality objectives are in this respect customer orientation or orientation towards the consumer and chain orientation. The latter means to recognise oneself within the value process as a customer and also as a supplier and to see the consumer at the end of the chain as a joint customer. Intercompany quality policy is used to establish a joint philosophy and way of thinking. Intercompany QM systems frequently come up against company limits, limits in the area of competence and responsibility of the actors involved in the value chain, country limits and personal and financial limits. Only with suitable organisational structures for coordinating internal and intercompany QM tasks and by professionally managing the supplier relationships, these limits can be overcome. Despite this, product warning calls and targeted product recalls are owing to their health risks time and time again events requiring intercompany risk management. Innovative solutions to design risk management and for early detection of national and international risks can only be developed through concerted action of all actors involved.

DOI 10.3920/978-90-8686-789-9_8, © Wageningen Academic Publishers 2014

8.1 Quality objectives and quality policy

B. Petersen[1] and V. Schütz[2]
[1]*University of Bonn, Katzenburgweg 7-9, 53115 Bonn, Germany*
[2]*DRV, Deutscher Raiffeisenverband e.V., Pariser Platz 3, 10117 Berlin, Germany*
b-petersen@uni-bonn.de

Relationships between customers and suppliers in the agri-food industry are very complex and heterogeneous. Just as diverse are therefore the possibilities to design intercompany QM systems. However, a driver is always necessary to initiate the establishment of corresponding management systems. Any obstacles appearing in this respect can be overcome by consensus regarding achievable advantages and synergies as overall joint objectives. The important findings from the experiences made in the past years are as follows: by improving the own quality management, each organisation involved can at the same time increase efficiency of the quality management of other organisations participating in the value chain, and this will create trust amongst them (Schulze Althoff, 2006). Thus, in times of crisis, those value chains which have established sustainable cooperations over various production stages within the framework of joint QM strategies can react more quickly and more efficiently.

The providers of agricultural products and foodstuff have recognised since the mid 1990s clearly that they have to attract the attention of customers more and more intensively. The quality of a product is not the only decisive criterion when buying but is rather a matter of course. Companies of the agricultural producer stage and food processors have since then acquired multiple experiences as to how customers can be gained and kept and how they should react to which customer requests without delay. Therefore, it is nowadays of utmost importance to already anticipate customer requests and developments on the market ahead of others and integrate them with the greatest speed possible in product and process development.

A value chain as organisation needs just as much as an individual company a role model and a philosophy underlying the QM system. The most important quality objectives are in this respect customer orientation or orientation towards the consumer and chain orientation. The latter meaning to understand oneself within the value process as a customer as well as a supplier and to view the consumer at the end of the chain as a common customer. In this regard, coordination and consensus on measures successful in aligning business performance with the needs and demands of external and internal customers of a production chain are important. In the foreground is the question of how the entire chain from agricultural production to food retail can be included.

The main focus of an intercompany quality policy is on establishing a common philosophy and way of thinking. This means to define in a first step 'what' is wanted and then to plan 'how' this can be realised. When answering the question 'how', it is quite often possible to refer to progress reports from other industries which can be used. This applies to intercompany methods as well as to strategies (see Chapter 2.2).

Structuring of quality policies according to philosophy, strategy and methods is to be understood as a kind of modular system (Figure 8.1). In order to reach the next higher level in each case, i.e. from the method to the strategy or from the strategy to the quality philosophy, several elements of the subordinated level form the basis for proceeding systematically.

According to this allocation, international standards, such as ISO standards 22000 or 9000 (see Chapter 5.1), are to be seen as strategies since they contain requirements necessary for organisational implementation of the quality philosophy. These include for example:

- identification and traceability to create transparency within the chain;
- supplier assessment and supplier promotion to be able to pass on quality guarantees to the next customer within the association or value chain;
- preventive and continuous improvement measures to support a lively and agile QM system;

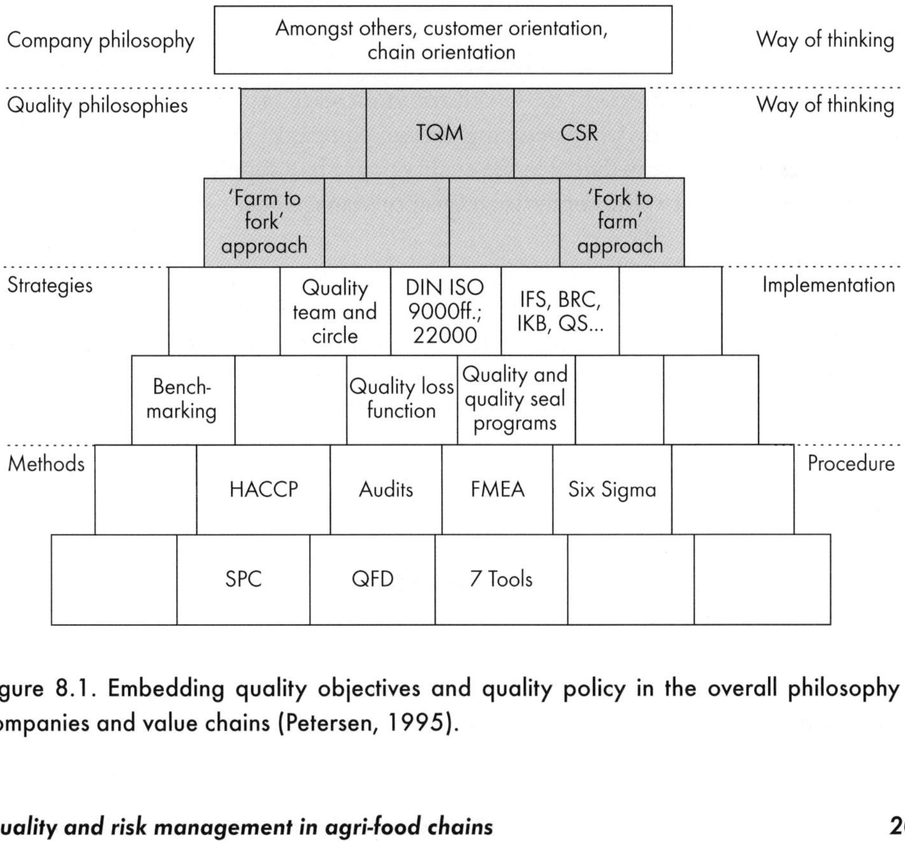

Figure 8.1. Embedding quality objectives and quality policy in the overall philosophy of companies and value chains (Petersen, 1995).

- creating contractual and legal security between business partners and towards third parties by means of documentation and corresponding verification.

Quality policies include all quality-related principles for guidance and action and thus form the framework for all QM activities. In this connection, two overall principles are 'farm to fork' and 'fork to farm'. Mostly it depends on the type of driver within the chain which of these principles is implemented in an intercompany quality policy.

With the 'farm to fork' principle, the initiative to design an intercompany QM system most of the time starts at the agricultural production stage. Often farmers organised in cooperatives are not only defining the quality objectives and thus the quality policy in primary production but also in the processing stages. This includes all measures taken within the scope of product design as part of the political instruments used for sales and procurement. Quality policies are used in this process preferably as a competitive means together with other value or supply chains to food retail.

In contrast to price policy, quality policy promises a competitive advantage which competing chains cannot make up immediately but only after a certain time. Delivery reliability, safety of products, transparency and traceability guarantee as well as flexibility are often amongst those statements regarding quality policy which are translated into measurable quality objectives and subjected to continuous internal and external audits.

With the 'fork to farm' approach, the last distribution stage (e.g. retail) has to be seen as the driving force in the chain when designing an intercompany QM system. Depending on how many stakeholders are included in the quality policy by a supply chain, a distinction is made between the quality philosophy Total Quality Management (TQM) and Corporate Social Responsibility (CSR).

TQM (Chapter 2.2) is oriented towards the process of production, processing and trading with feedstuff and foodstuff. This principle is founded on the conviction that quality is a question of orientation towards the requirements of customers and that these customer requests are measurable.

The TQM strategy is characterised by six features:
- the quality of all management, business and administration processes between the individual stages of the chain;
- a culture of continuous improvement of all companies within the chain;
- the understanding that quality improvement leads to cost benefits and an improved profit potential;
- the design of intensive cooperation between customers and suppliers or other interest groups involved in the value process;

- including all actors of the chain;
- market orientation of all departments and business units in the case of individual companies.

The initiative to introduce TQM in entire value chains is often taken when groups of customers and/or consumers suggest how and which products are to be produced and delivered in future. Quite often it is the food retail actively approaching delivery chains and demanding implementation concepts. Typical objectives to achieve quality leadership lie according to experiences made in the areas of food safety, animal protection, environmental protection, sustainability, etc. So far, only few supply chains have been successful in putting all of the six above mentioned TQM principles into practice in an intercompany QM system.

Weak points often encountered are the following, i.e.:
- insufficient commitment of the individual stages of the chain and its management;
- unrealistic expectations;
- lacking determination of priorities;
- inadequate measuring methods.

Management within the TQM concept means shifting the main focus, i.e. away from the rigid form of system administration and towards genuine leading targeted to finding a consensus in the chain and a future version. One of these visions might include a stronger orientation of the quality policy towards sustainable development exceeding legal requirements (compliance).

Corporate Social Responsibility (CSR) is synonymous with acting responsibly. An intensive dialog with various stakeholders can support value chain to orient their commitment in the field of CSR in a targeted manner towards the needs and requirements of their stakeholders. In addition to this, this type of communication strengthens motivation of companies to get more involved in the area of social responsibility. Experience from the industry shows that the extent of dialog amongst stakeholders positively influences the scope of CSR activities (Hartmann, 2011).

Sustainability has become in recent years in this connection a guiding theme in the agri-food industry. Already for some time now, the strategic objective of sustainability and the related measures have been determining the economic corporate activities in the value chain (Brinkmann and Petersen, 2011). This is due to the fact that the agri-food industry is confronted with the task to double agricultural production up to the year 2050 owing to the growth of population from now 7 billion to up to then 9 billion people. This increases the pressure on the use of resources such as raw materials, land and biodiversity. This can be done only if organisations are successful in the area of tension between ecology, economics and society to achieve higher yields with less resources, maintain profits for agricultural enterprises, protect the environment and fulfil social requirements (Breitschuh and

Gerhard, 2012). Therefore, sustainability aspects will play an important role for the strategic orientation of all organisations in the industry in future. At the moment, key criteria how to determine a status of sustainability are considered for various concepts and models (Table 8.1). In some areas, these criteria are different for vegetal and livestock production.

Table 8.1. Criteria to categorise sustainability of production branches of agricultural enterprises (Breitschuh and Gerhard, 2012; Klauke, 2012).

Main criteria	Partial criteria	
	Vegetal production	Livestock production
Ecology	Soil: humus balance, nutritional content, compression, erosion, biological diversity status, agricultural environmental measures, protected areas, ecological toxicity, intensity, nitrogen surplus, crop rotation, outcrossing potential	Animal populations: genetic resources, races, feed consumption, animal health, liquid and solid manure, farm balances regarding nitrogen and phosphor
	Farming: area under cultivation, type of soil cultivation and crop rotation	Use of buildings: rearing and feeding systems, storing of feedstuff
	Emissions: air, greenhouse gases, acidification, depletion of the ozone layer, photo-chemical oxidation, sewage, solid waste, liquid and solid manure	
	Water: consumption	
	Resources: abiotic consumption	
	Energy consumption: non-renewable energies, renewable energies	
	Ecological toxicity: risk potential	
Economy	Variable cost: soil cultivation, seeds, plant protection, fertilisation, machinery	Variable cost: feedstuff, veterinary drugs, consulting, meat quality, meat safety
	Fixed cost: depreciation, maintenance, insurances, labour cost, investments, other fixed cost	
	Macro economy: subsidies, productivity, gross value creation, further economic factors	
Social matters and society	Consumer: safety and quality of vegetal products such as cereals, fruits, vegetables, wine and ancillary products; residues, GMOs in foodstuff, toxicity, product characteristics, other risks	Consumer: safety and quality of animal products such as meat, milk, fish, eggs and ancillary products; drug residues, rearing conditions, animal welfare
	Employees in companies: wages and salaries of permanent employees and seasonal workers, social contributions, further education, membership in associations, representation of interests, wages of suppliers, risk potential at work, compatibility of family and job, pension schemes	
	Local and national community: regional production, access to land, employment, equal rights, integration, qualification of workers, rural development, urban-rural dialog, information platforms and exchange of experience (www.drv.de)	
	International community: import from developing countries, fair trade, foreign direct investments	
	Future generations: involvement in professional formation, expenditure for research and education, investments	

Box 8.1. Sustainability: origin and importance.

The term 'Sustainability' comes from forestry management and was first used in 1560 in the Saxon Forestry Order. The term became known because the World Commission on Environment and Development (Brundtland Commission) instituted by the United Nations developed a concept for 'sustainable development'. Since this time, politics and industry speak of sustainability. The final document which became known as the Brundtland Report 'Our common future' from the year 1987 defines the concept of 'sustainability' as follows: 'sustainable development is development that meets the needs of the present without compromising the ability of future generations to meet their own needs.' (WCED, 1987: 43)

In socio-political terms, sustainability is comprehensively understood to be 'current thinking and acting under the aspect of improving the living conditions of the generation of today without deteriorating the future perspectives of the generations of tomorrow' (Brinkmann and Petersen, 2011).

An increasing number of consumers expect from producers involved in the value chain of meat, milk, fruits and vegetables and cereals, statements regarding the above mentioned criteria and hence on their intention to act responsibly towards society and the environment. Aspects of sustainability therefore gain more and more importance for the purchase decision of customers and their willingness to pay (Hartmann, 2011; Spiller, 2010). In this way, new extrinsic characteristics regarding the quality of agricultural products develop. Companies can use such aspects when determining their quality policy to distinguish their own products in existing markets, exploit new market segments and to improve the economic situation of all those involved in the value chain. This can only be achieved if all actors of the value chain work together, and in this respect agricultural primary production typically plays the most important role.

If the joint quality policy has been determined, it is necessary to define strategic and operative quality objectives in each stage of the chain and determine methods to measure them. Typical quality objectives are for example increasing customer satisfaction to be measured by the proportion of regular customers and new customers during a defined period of time; or employee motivation to be proven by the concrete number of days spent on further education or by additional competences shown by employees; or by the number of internal audits carried out.

Box 8.2. SMART rule.

Quality objectives serve to implement quality policies and define the benchmarks for quality targeted towards the inside and towards the outside within a company. It is necessary that they can be measured and characterised by means of the SMART rule:

S = specific: A clear and well defined partial goal can be set.

M = measurable: The degree of success can be observed or measured indirectly.

A = acceptable: a consensus can be achieved that this goal is to be pursued (in addition to other goals).

R = realistic: the goal can be reached taking existing financial, personal, political and other framework conditions into account.

T = time-bound: a deadline for achieving the objective is defined.

8.2 Organisational structures of intercompany quality management

V. Schütz[1] and B. Petersen[2]
[1]DRV, Deutscher Raiffeisenverband e.V., Pariser Platz 3, 10117 Berlin, Germany
[2]University of Bonn, Katzenburgweg 7-9, 53115 Bonn, Germany
schuetz@drv.raiffeisen.de

Basically there are four limits for the actors of a value chain from agricultural enterprises up to food retail or canteen kitchens (last distribution stage) planning to introduce an intercompany QM system (Petersen, 2003; Schulze Althoff, 2006):

- company limits since QM systems are initially often only oriented towards individual enterprises;
- limits in the area of competence and responsibility of the actors of the value chain;
- country limits characterised by different languages, production conditions, quality and information standards;
- personal and financial limits when organising intercompany information and communication systems.

Against this background, the challenges mainly consist in aligning and developing suitable organisational structures for coordinating internal and intercompany QM tasks.

Differences of existing organisational structures between individual companies of the agri-food industry can be found, on the one hand, regarding the intensity of contractual relationships of the participating actors (degree of loyalty), and, on the other hand, in respect of the powers of the relevant actors when determining requirements (degree of centralisation) (Grossekettler, 1978; Trienekens and Beulens, 2001a; Schramm and Spiller, 2003; Spiller *et al.*, 2005).

Thus, categorisation can be carried out by answering the following questions:

- Where can the decision competence be found?
- Which actor initiates which type of measures?

In practice, the various forms of contract are considerably different regarding their degree of loyalty and centralisation (Drescher, 1993). Kagerhuber and Kühl (2002) describe the individually concluded contractual relationships along an 'integration continuum'. The degrees of loyalty and centralisation continually increase between the categories of spot market, informal relationship, cooperation and company hierarchy (Table 8.2).

Following joint consideration of horizontal and vertical relationships in the Netchain (see Chapter 4.1), the forms presented here can be applied horizontally as well. As an example

Table 8.2. Contract criteria on the basis of the integration continuum (Schulze and Althoff, 2006 from Kagerhuber and Kuehl, 2002).

Criteria	Spot market	Informal relationship	Cooperation		Hierarchy	
			Synergetic system	Partners	Hierarchical system	Fusion
Description	• Ad-hoc coordination • No planned organisation of relationships as a precautionary measure	• Anonymity is no longer possible amongst the market participants • Adaption of products towards the preferences of the market partners at a certain basic level • Adaptation of products regarding preferences of market partners (at a low level)	• Organisational relationship without making use of asymmetrical market power • Non-cooperative behaviour of the market participants is possible and dangerous for the system	• Interlinking of capital or reduction of decision rights for the own company by concluding contracts • Non-cooperative actions are restricted de jure and conditionally dangerous for the system	• A priori asymmetrical distribution of power • Unilateral, non-cooperative behaviour is only possible to a limited extent and cannot destroy the system	• Dissolution of original systems • Complete identity taken into responsibility
Duration of contract	short-term	long-term				
Interaction between...	unknown participants	known participants			not relevant	
Type of coordination	ad hoc	planned				
Distribution of decision-making rights / market power	symmetrical	symmetrical	symmetrical (in the optimum case)	weighted according to proportionate participation	asymmetrical	only one party
Degree of integration	marginal integration		partial integration		complete integration	

for a horizontal form of cooperation with specific contractual characteristics, a producer association or an animal husbandry cooperative can be mentioned (Bassewitz, 2002; Karge, 2002; Spiller *et al.*, 2005). Depending on the degree of loyalty and centralisation, a broad spectrum of economic and social objectives can be pursued with regard to the joint use of production capacities (Doluschitz, 2001; Schulze Althoff, 2006; Spiller *et al.*, 2005).

The requirements existing within a QM system refer to the own organisation, on the one hand, and to the relevant suppliers, on the other hand. Figure 8.2 explains this relationship. The more customer-supplier relationships with its many agreements are considered in a QM system, the more complex the tasks become and the stronger grows the demand for coordinating support measures. Therefore, in the past few years new service fields of so-called 'bundler organisations' and 'network coordinators' have developed (Düsseldorf *et al.*, 2011; Ellebrecht *et al.*, 2009; Schütz, 2009).

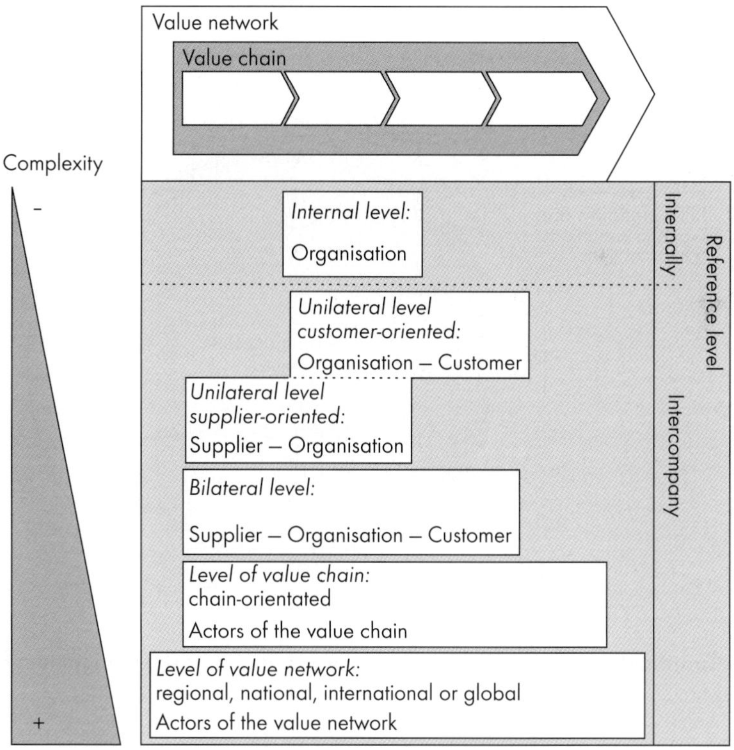

Figure 8.2. Reference levels of quality management.

Network coordinators

Network coordinators take over intercompany services along the value chain mainly in three different areas:

1. Strategic and operative planning in quality and risk management of individual companies in the chain and customer relationship management.
2. Standardisation of audit, document and measure management as components of a service process.
3. Creation of technical and organisational interfaces for communication within the chain.

Network coordinators exclusively fulfil service functions and organise development and structures for intercompany cooperation within the framework of quality management (Figure 8.3). Furthermore, they offer support for process management, audit management, supplier management and crisis management, of each individual enterprise and thus at the same time for the entire quality chain or quality network.

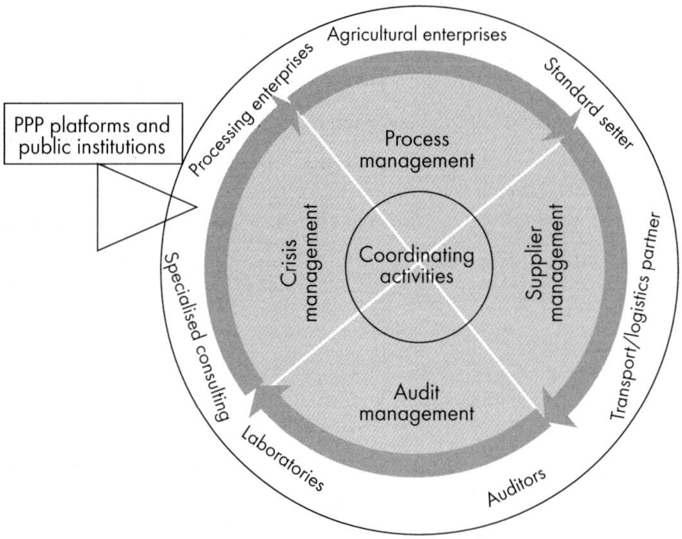

Figure 8.3. Coordinating services in intercompany quality and risk management for different stakeholders.

Bundler organisations and categories of intercompany QM systems

Network coordinators can also be bundler organisations in the sense of national quality seal programs. These developed within the framework of initiatives to assure traceability and quality of products taken throughout Europe in the past few years by different actors in particular of the value chain meat, milk and fruits and vegetables. In Europe, in particular three models of procedure can be distinguished to implement a uniform quality policy and to coordinate quality management tasks: open, closed and mixed chain-related quality program (Table 8.3).

In the 'open system', an external program owner defines the criteria and requirements placed on a quality and risk management system for the individual production stages nationally for a chain. Referred to the meat-producing chain, already by the end of 1990, concepts

Table 8.3. Criteria of different forms of intercompany systems in quality and health management (modified from Schulze Althoff, 2006).

System categories	Criteria
Closed system	Coordinated organisation of quality management of the partners of a Netchain and: • cooperative behaviour of system participants (e.g. brand meat programs) • own requirements on system participants, e.g. on production, rearing of animals, cultivation methods • value creation up to the counter or consumer is possible • system is product and market-specific • QM system is supported by own coordinating institution • internal control systematics including external auditors • contractual requirements are determined internally • long-term involvement of system participants by contributed capital
Open system	Requirements on quality assurance and type of conformity tests are primarily determined by external systems (e.g. IKB, QS): • sanctions for not adhering to contractual requirements are determined externally • few requirement criteria exceeding legally controlled areas • often without actively communicating the approach of a continuous improvement • system is oriented supra-regionally, mostly nationally
Mixed system	Combination of elements of the closed and open system: • in addition to participation in an open system, within chain-internal customer-supplier relationships advanced quality standards are processed • intercompany coordination between production stages in quality management is aligned • open systems are used as the basis for establishing bundler organisations and sector-specific audit activities

were developed first in Denmark and the Netherlands which determined national quality programs for supplies to retailers.

In the 'closed system', the concrete market partners agree within the existing fixed customer-supplier relationship of a chain a joint quality policy and determine in all individual contracts amongst them the requirements for a joint QM system. Another organisational strategy presents the 'mixed system' where the basic requirements of open systems are supplemented by own chain-oriented quality systems.

Communication via network coordinators

Successful quality management requires intra-organisational and inter-organisational communication. Communication between the hierarchical structures of an organisation is just as important as communication between the organisations involved in the value process. With increasing heterogeneity of a network, access to the network (Schenk, 1995) and thus to quality management can be increased by information and communication technologies. With this, data management – including relevant information and communication services – becomes more and more important in value chain, not least owing to the increased quality requirements according to the legal requirements of the EU Hygiene Package (Schütz, 2009).

The requirements placed on communication of figures, data and facts during the production process, rise with the complexity of value networks. To communicate information relevant for quality management with time, information and communication structures (IaC) especially in regional networks developed. From commonalities and differences of web-based information platforms serving at least two or more network levels, development potentials can be shown by way of benchmarks.

Seen retrospectively, two development drivers often initiate regional networks:
1. safeguarding sales volumes with the objective to establish regional brands;
2. building up intercompany quality and risk management systems with a focus on introducing performance controls during production and monitoring procedures.

A current trend is seen in regional organisational structures of networks uniting both development approaches. Most of the information platforms have a regional radius, but supra-regional web-based information platforms have developed as well. Examples for supraregionality are the animal health agency (TiGA) or QS information and registration platforms (Nienhoff, 2008).

Quality communication is defined according to Meyer (2010) as 'the totality of all communication efforts to make noticeable and transparent measurable and non-measurable quality characteristics regarding communicated and non-communicated quality requirements for the relevant stakeholders'. Transmission of information to the food chain according to the procedure of 'one step up, one step down', is legally required to increase transparency of goods streams or animal movements (Theuvsen, 2003). The EU Basic Regulation (EC) 178/2002, the German Livestock Movement Order (VVVO) and the EC Animal Transport Regulation are amongst them and back simultaneously the establishment of IT-supported IaC systems. In a crisis situation, goods streams can be traced back quickly (aid Infodienst Verbraucherschutz, Ernährung, Landwirtschaft, 2003). The Basic Regulation (EC) 853/2004 (EU, 2004b) 'Specific hygiene rules for food of animal origin' demands communication along the production chain (Meemken and Blaha, 2008). The Basic Regulation (EC) 854/2004 (EU, 2004c) 'Specific rules for the organisation of official controls on products of animal origin intended for human consumption' has created the framework conditions for implementation of risk-oriented inspection strategies for example between the slaughtering stage and the suppliers. By the cross-stage exchange of additional information on rearing, animal health, epidemiologic data and foodstuff safety, the inspection frequency at the slaughtering belt can be varied depending on the risk involved.

Regarding quality communication, two different types of communication can be distinguished in general: communication between supply chain and end consumer at the point-of-sale takes place via labels on the packaged end product, such as the QS certification mark, a seal to identify organic production, the ProPlanet label or the indication of origin 'Tested Quality Bavaria', etc. (Plumeyer *et al.*, 2008). The second form represents communication between customers and suppliers within the value chain. The exchange of results from product and process audits by means of figures, data and facts within the value networks is in the foreground here (D. Brinkmann, personal communication).

Assessment of coordination services

Some literature contains various suggestions on the criteria according to which these origin securing systems and brand programs can be sorted regarding consumer requirements or acceptance (D. Brinkmann, personal communication). In this context, the aspects of regional marketing are in the foreground and not the observation of intercompany coordination tasks or the technical and organisational side of quality communication in the network. However, literature also contains various approaches for the assessment of the advantage of intercompany information and communication systems and interaction between service users and service providers. Accordingly, Schulze Althoff (2006), Mack (2007), Ellebrecht (2008) and Schütz (2009) elaborated partial assessment approaches which are summarised in Table 8.4. While the work of Schulze Althoff (2006) and Ellebrecht (2008) strongly refer to the technical side of quality communication – information and communication systems, Mack (2007) and Schütz (2009) emphasise in their planning models more strongly the organisational aspects of a network coordinator. On the basis of this methodical

Table 8.4. Overview of partial assessment approaches and model parameters with reference to intercompany quality communication in regional networks (Düsseldorf *et al.*, 2011).

Type of model	Model parameters	Authors
6-stage model to design quality communication	1. decision criteria for designing the IaC system 2. degree of organisation 3. time necessary for introduction and further development 4. smallest traceable unit 5. extent of data input in the IaC system 6. type of processing and transformation of data into information with business intelligence tools	Schulze Althoff, 2006
Value index for intercompany information and communication systems	$\text{Value index} = \frac{ZG\,(1+\Delta d) \times IZ \times \ddot{U}G}{ZE\,(d)}$ ZG = time gain in days IZ = information gain ÜG = degree of similarity in data exchange ZE = period for decisions	Ellebrecht, 2008
Service index for benchmarking network coordinators	DI = service index di = service intensity dk = service complexity a = audit type combination d = document type combination m = amount of service users per year f = functional combination h = audit frequency per service user and year	Mack, 2007; Schütz, 2009

approach, a proposal for characterising quality communication in regional networks has been developed. Five coordinating service providers selected from a population of 102 QS bundler organisations served as research objects.

The assessment models for value networks of the pork-producing chain have been developed by way of example. The example serves to describe the basic principle of intercompany systems to support quality communication. Two benchmarks stood in the foreground of assessment:

1. Effective coordination regarding different inspection activities (system, process and product inspections) (types of audit).
2. Coverage within the value network and thus type and number of different stakeholders within the system.

Here, it was verified to which extent and in which way seven benchmarks were fulfilled by the organisations. For this, altogether 44 evaluation criteria were defined (Table 8.5). For each benchmark, a degree of fulfilment amounting to 100% was always assumed if all the criteria assigned to the benchmark were actually demanded by the network coordinators as a service offer in practice and were also made available.

The results of evaluation for five research objects are summarised by way of example subsequently in the form of a radar plot. In condensed form, from this typical strengths

Table 8.5. Evaluation criteria (Düsseldorf *et al.*, 2011).

Benchmark	Criteria	
1. Actors [1,2,4]	1. Breeder	8. Consultant
	2. Piglet producer	9. Farm veterinary
	3. Fattener	10. Official veterinary
	4. Slaughterhouse	11. Forwarder
	5. Food retail	12. Laboratories
	6. Producer association	13. Feedstuff producer
	7. Animal trade	
2. Supplier management [2,4] (activities)	1. Assessment of food chain information	4. Forward and backward information of process and product codes
	2. Forward and backward information for risk-oriented meat inspections	5. Reports on supplier assessment
	3. Planning of purchase and sale	6. Daily report of slaughtering data
3. Crisis management [2,4] (activities)	1. Notification of authorities in a crisis	3. Multiplication of authority information
	2. Document control in crisis situations	
4. Process management [2,4] (activities)	1. Operational branch evaluation	3. Notification of animal transports
	2. Comparative enterprise evaluation	4. Process documentation
5. Audit management [3,4] (activities)	1. Audit planning, rectification and result storing	6. Integration and processing of data from the audit management system
	2. Internal and external procedural audits	
	3. Product audits	7. Audit type veterinary stock supervision
	4. Document control	8. Audit type internal technical production audit
	5. Audit protocol and result	9. Audit type QS certification audit
6. Audit groups (types of audits)	1. System audits	3. Product audits
	2. Process audits	
7. Coverage (network levels)	1. Breeding	4. Meat centres
	2. Piglet production	5. Trade
	3. Fattening	

The criteria are based on: [1]Schulze Althoff (2006); [2]Ellebrecht (2008); [3]Mack (2007); [4]Schütz (2009).

and weaknesses of the organisation under surveillance can be deduced (Figure 8.4) A radar plot consists of seven axes according to the individual benchmark selected.

The regional network (F1) consisting of a cooperation between producer association and food retail shows the highest coverage, as could be expected. One reason for this is in this case the joint responsibility for a brand meat program. All value stages must be integrated to secure raw materials and to implement a targeted marketing campaign at the point of sale effectively. Therefore, another cooperation with a joint quality policy between producer and meat processing stage exists and the last distribution stage being the food trade, i.e. butchers, achieves a coverage just as high (F3). The assessment criteria 'integrated actors' shows a very heterogeneous image. The results vary from a degree of fulfilment of 45% to 90%. Above all, the network mainly active as QS bundlers (F4) serves less actors than networks acting in cooperation with trade and craft (F1 and F3). All of the five networks achieve the highest degree of fulfilment of 100% with the criteria in the area of 'supply management'. Referred to the benchmark of 'crisis management', significant differences can be seen regarding the degree of fulfilment between the two networks considered (0 up to 100%). Process management is almost completely fulfilled by all networks (exception F2: 25%). With the benchmark of 'audit management', the network (F3) cooperating with food trade stands out positively. All networks show potentials for optimisation in the area of 'audit types', i.e. supporting additional types of audit could become a further service field of network coordinators.

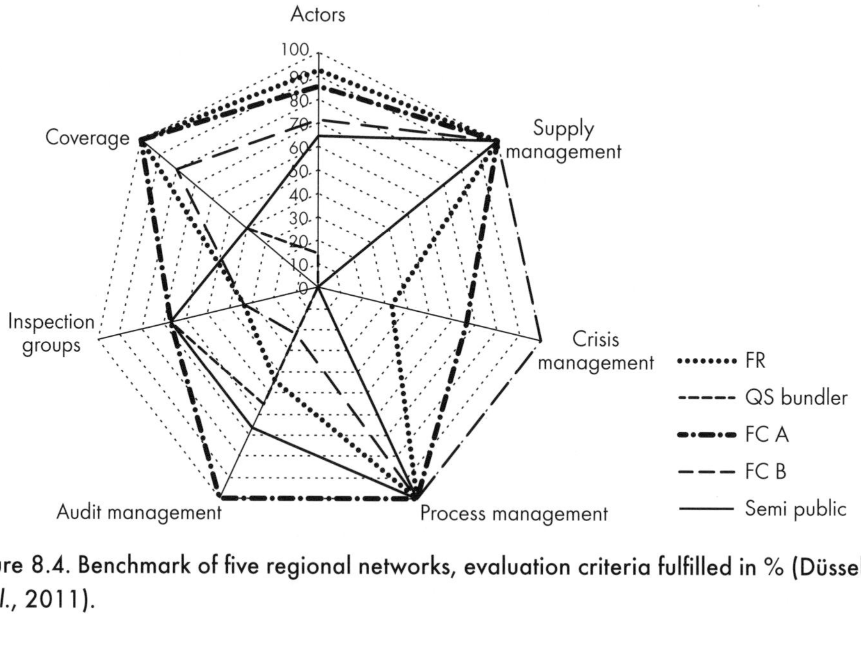

Figure 8.4. Benchmark of five regional networks, evaluation criteria fulfilled in % (Düsseldorf *et al.*, 2011).

The above mentioned seven benchmarks are suitable to illustrate the regionally and organisationally different networks with their specific profiles on a comparative basis (Düsseldorf *et al.*, 2011). There is a considerable need for research, however, to develop a methodical procedure for including further benchmarks considering qualitative and quantitative criteria. A promising methodology is the assessment of the degree of maturity according to ISO/IEC 15504. In contrast to an audit where conformity is verified with certain requirements, determination of the degree of maturity according to ISO/IEC 15504 also places particular emphasis on the capacities for fulfilling these requirements (Schmelzer and Sesselmann, 2008). Contrary to benchmarking, instead of using the processes of other companies as a comparison, the assessment is rather based on the best condition possible. For this, elementary procedures in an abstract process model are registered, structured and illustrated. Such a model having a recommendatory character for reuse, is usually called a reference model (Brenner *et al.*, 2010). Determining the degree of maturity by means of a reference model is an approach which has been selected increasingly in recent years (Wagner and Dürr, 2008).

Models to determine the degree of maturity together with the assessment of processes form an important step towards professionalization of companies or value chain, especially in software and system development. The success of maturity models makes it possible that nowadays the benefits of this procedure can be used across industries. The first step here is to subdivide the maturity of processes into maturity stages (Köhler, 2006). By means of the objective to achieve the next higher degree of maturity, subsequently the strengths and weaknesses of a process can be uncovered. This facilitates targeted process improvement. There are several reasons for continuous process improvement. One very important reason is maintaining or improving customer satisfaction. This is decisively determined by the fact how the customer, for example, perceives the quality of a service offered by a network coordinator. All weaknesses and faults discovered by the customer affect customer satisfaction and long-term customer loyalty negatively (Slütter *et al.*, 2010). Indices designed by Mack (2007) for rating service intensity and complexity offer another alternative for extended benchmarking.

When evaluating the results, it has to be made sure that benchmarking is based on a status-quo survey. Most of the networks are geared towards an extension of their service portfolio by adding modules. One possible extension module is, for example, the monitoring and certification procedure introduced in connection with the Tiergesundheitsagentur eG (TiGA). It would also be conceivable to include modules allowing dynamic management of sustainability aspects along the value chain.

Ellebrecht (2012) developed a concept for efficiently designing services in intercompany health and quality management by means of team-oriented QM methods. In this connection, the author wanted to find an answer to the question to what extent existing service offers still correspond with the actual customer expectations of farmers. In addition to this, she determined the level of awareness of QM methods with network coordinators of the agri-food industry. Answering these investigative questions was the precondition for assessing the development potentials of future service offers with the support of team-oriented QM methods. After completing data collection, a consulting and assessment concept for the extension of the service portfolio offered by network coordinators was presented. In a diagram consisting of four quadrants, the questioned 22 network coordinators from three EU member states could be subdivided according to their service offer in intercompany health management (CS) and the use of QM methods (Figure 8.5). Five organisations belong definitely to the 'stars'.

Ellebrecht's application concept supports development of network coordinators to become 'stars' including both, i.e. method-supported quality management and coordination of services in the areas of product consulting, supplier assessment, crisis management and audit management when using QM methods. In contrast to this, the modified Kano categorisation enables to optimally consider customer expectations when extending

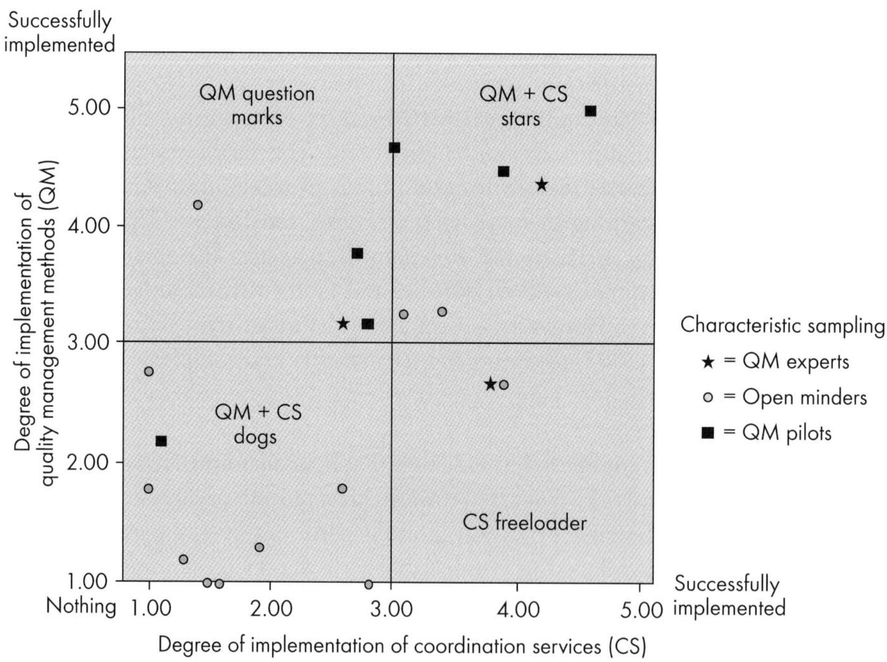

Figure 8.5. Assessment diagram for categorising network coordinators according to the degree of implementation of coordinating services and quality management methods (Ellebrecht, 2012).

service portfolios. For example, services not demanded directly by farmers but creating an enormous added value for them are classified as 'attractive services' (Ellebrecht, 2012).

The procedures and methods for contrasting, assessing and further developing of services offered by network coordinators were explained in this chapter predominantly by way of examples from meat production. The results and experiences can basically be transferred to other value chain of the agri-food industry where network coordinators take over certain tasks.

8.3 Supplier assessment and supplier promotion

L. Theuvsen
Georg-August-University Göttingen, Department of Agricultural Economics and Rural Development, Platz der Göttinger Sieben 5, 37073 Göttingen, Germany
ltheuvs@gwdg.de

In the agro-industrial sector predominantly characterised by medium-sized enterprises, procurement management has not been in the focus of attention for quite a long time. This situation has changed fundamentally. Strategic orientation of procurement and professional management of supplier relationships have drastically increased in importance (Heyder *et al.*, 2009). Various reasons are decisive for this. Little by little, the great importance of supplier management has been recognised, amongst others, in connection with assuring competitiveness of companies, improvement of relationships with suppliers to safeguard the raw-material basis (Wocken and Spiller, 2009), traceability of products, transparency of value chain (Deimel *et al.*, 2008a) and quality assurance.

Procurement serves economic and needs-oriented supply of companies with input factors (raw materials, packaging material, etc.). A central component of the corporate procurement process is supplier management (Figure 8.6). It encompasses seven partial processes (Arnold, 1997):
- supplier identification;
- supplier preselection;
- supplier analysis;
- supplier assessment;
- supplier selection;
- supplier controlling;
- control of supplier relationships.

Within the framework of supplier assessment, the performance of suppliers is made transparent. Identification, assessment and selection of efficient and reliable suppliers has a central importance for purchasing and product quality (Bahlmann and Spiller, 2008; Hartmann, 2004). Thus, supplier assessment has an enormous levering effect on the success of a company.

Supplier assessment in the agri-food industry

How are companies of the agri-food industry selecting their suppliers? To answer this question, Heyder *et al.* (2009) have surveyed within the framework of an explorative study predominantly small and medium-sized enterprises in the German agri-food industry on the phone to find out about their procedures regarding supplier assessment. The participants

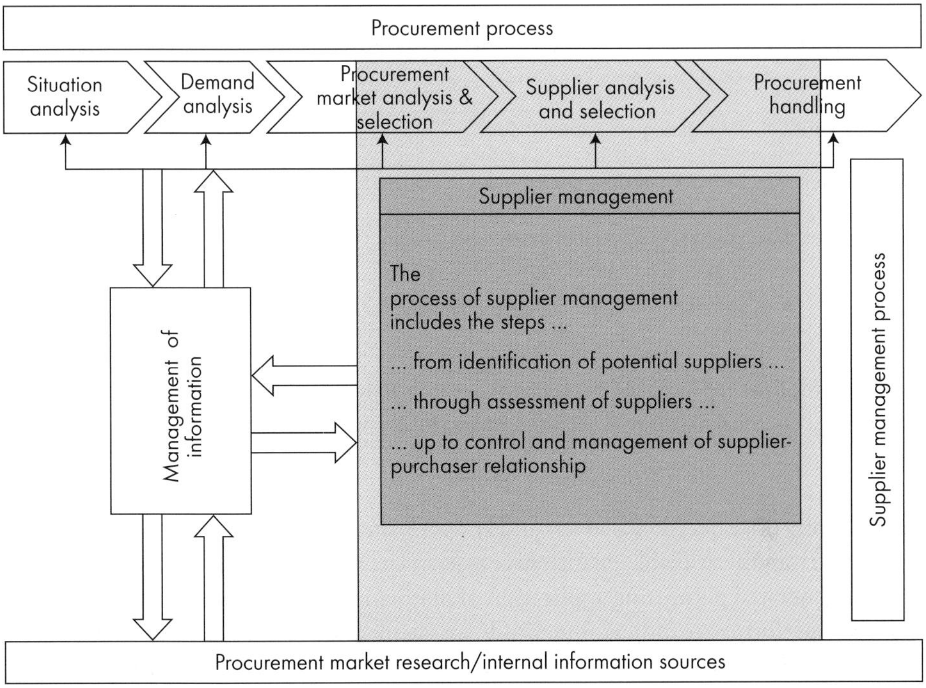

Figure 8.6. The procurement process (Arnold, 1997).

of this survey came mainly from the fields of pastry products and pasta (45 companies), dairy farming (33) and meat processing (20) as well as fruit and vegetable processing (16).

The survey showed that in the agri-food industry, the improvement of the quality of pre-products, ensuring safe supply and guaranteeing cheap procurement costs are the dominating reasons for carrying out systematic supplier assessment (Figure 8.7). Corresponding with the objectives mentioned, quality (μ=1.85; σ=0.4), price (μ=1.35; σ=0.7) and reliability (μ=1.20; σ=0.8) materialise as the most important criteria for the assessment of suppliers. Service (μ=0.86; σ=0.8) and general corporate characteristics (μ=0.53; σ=1.0) played a less important role. Table 8.6 illustrates what these criteria stand for individually in the agri-food industry and how important they are for supplier assessment.

Predominantly point rating methods, grading systems and checklists are used as qualitative assessment methods within the framework of supplier assessment (Figure 8.8). 94.5% of the companies are satisfied or even very satisfied with the checklist method used, 91.4% with the point rating system used by them, and 79.2% with their grading system.

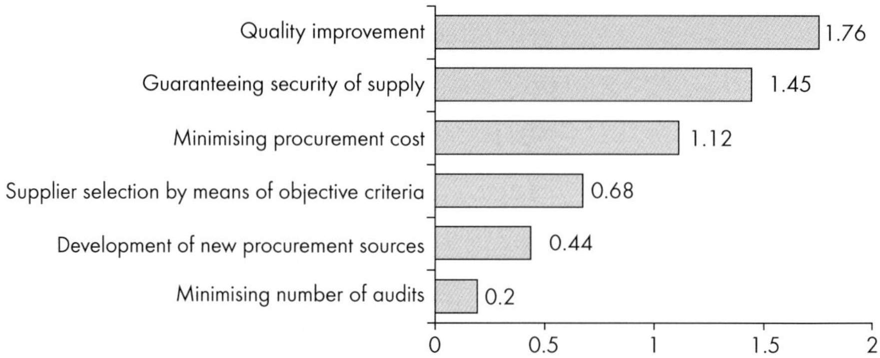

Figure 8.7. Motives for supplier assessment in the agri-food industry (mean values of a five-point Likert scale ranging from -2 = unimportant, to 2 = very important).

Predominantly price decision analyses and key performance indicator methods (Figure 8.9) are selected from the available quantitative assessment methods where figures or measured values are generated permitting application of mathematical or statistical methods. 70% of users of key performance indicator methods are satisfied or very satisfied with this method, while this applies only to about half of those carrying out a price decision analysis.

Complementary to the quantitative and qualitative methods mentioned, 86.2% of the companies questioned carry out a quality comparison, and 34.5% a delivery time comparison of suppliers. Frequently, more than one assessment method is used. 24.5% of the companies use a point rating method and a grading system. In 15.3% of the companies, a price decision analysis is carried out additionally, and in 10.2% of the companies a key performance indicator method is also used. 19% of the companies use both, i.e. point ratings and checklists.

Above all in the pastry and pasta industry, supplier assessment is carried out only once a year or even more seldomly. When procuring products with a high perishing risk, characteristic for milk, meat and fruit and vegetable processing, supplier assessment in contrast to this takes place in many companies more frequently (Figure 8.10).

Owing to the late shift towards supplier assessment, it is not really surprising that 61.9% of the questioned small and medium-sized companies still carry out supplier assessment without IT support. 11.9% of the companies use standard software, 22.9% software developed by them. The bigger the companies are, the more they rely on IT-supported supplier assessment (Figure 8.11).

Table 8.6. Importance of various criteria for supplier assessment (mean values of a five-point Likert scale ranging from-2 = unimportant, to 2 = very important).

		μ	s
Quality (μ=1.85, σ=0.4)	Importance of identified deficiencies	1.33	0.71
	Existence of certificates	1.32	0.82
	Complaint rate	1.24	0.83
	Quality management system	1.02	0.97
	Result of certification	0.82	1.01
Price (μ=1.35, σ=0.7)	Price level	1.42	0.80
	Price development	0.89	0.99
	Payment conditions	0.48	0.93
	Cost transparency	0.42	1.07
Service (μ=0.86, σ=0.8)	Flexible reaction to demand fluctuations	1.44	0.82
	Granted guarantees	0.58	1.04
	Customer and consulting service	0.38	1.02
	Goodwill services	0.38	0.92
Reliability (μ=1.20, σ=0.8)	Quality reliability	1.88	0.38
	Deadline reliability	1.63	0.71
	Price reliability	1.55	0.61
	Quantity reliability	1.38	0.68
Corporate characteristics (μ=0.53, σ=1.0)	Location	-0.13	0.87
	Employee qualification	-0.14	1.15
	Image	-0.22	0.84
	Available product spectrum	-0.26	1.19
	Size of company	-0.32	0.88
	Financial situation of supplier	-0.40	1.00
	Ownership	-0.60	0.84
	Profit situation of supplier	-0.76	0.87
	Employee fluctuation	-0.94	1.01
	Legal form	-1.03	0.77

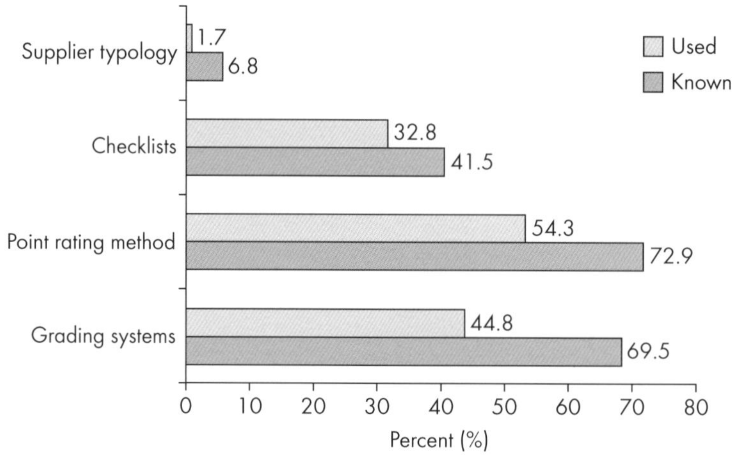

Figure 8.8. Knowledge and use of qualitative assessment methods (in %).

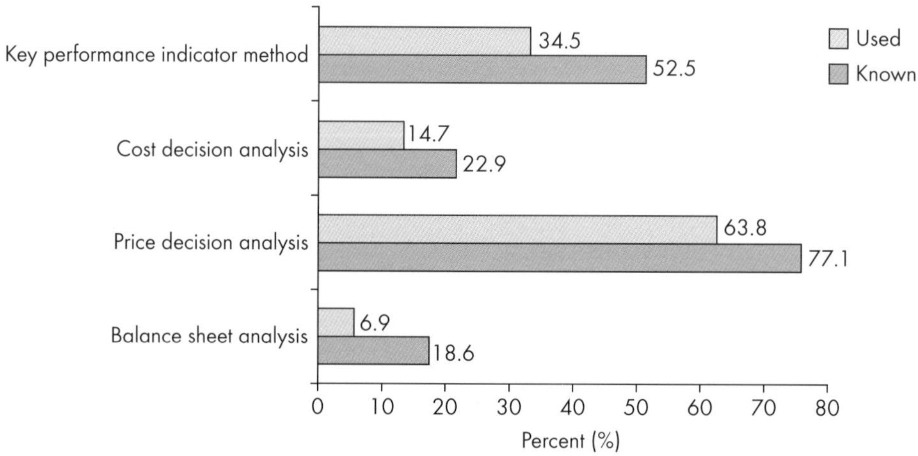

Figure 8.9. Knowledge and use of quantitative assessment methods (in %).

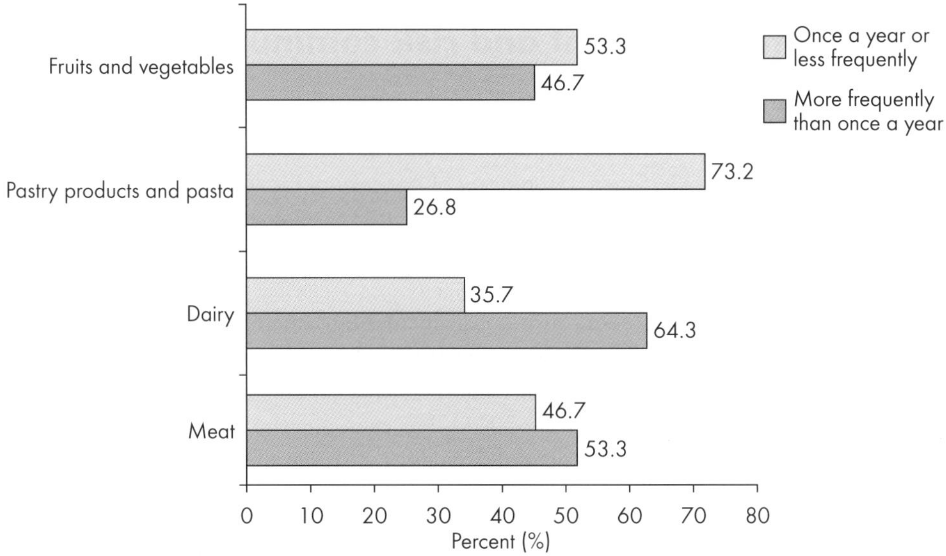

Figure 8.10. Frequency of supplier assessment (in %).

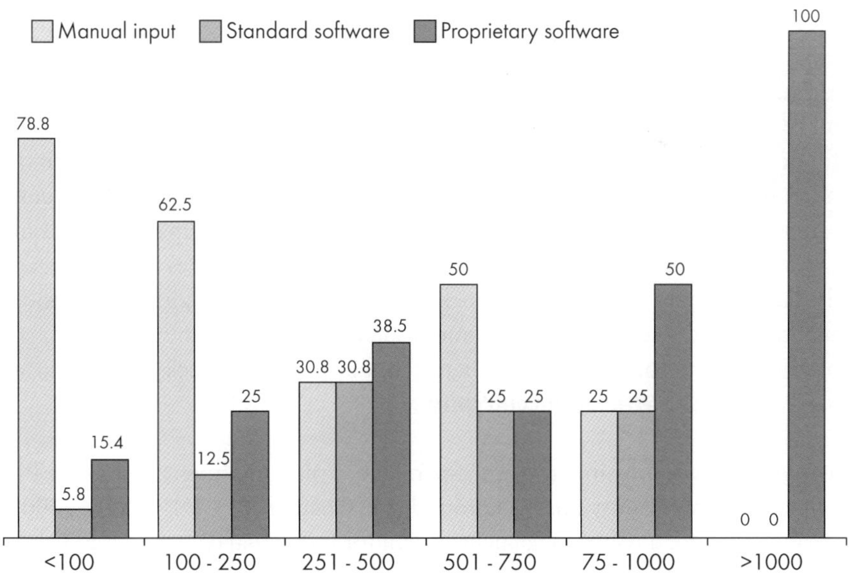

Figure 8.11. Use of IT-based methods for supplier assessment (in %).

8.4 Risk management and risk communication

V. Raab[1] and L. Theuvsen[2]
[1]DRV, Deutscher Raiffeisenverband e.V., Pariser Platz 3, 10117 Berlin, Germany
[2]Georg-August-University Göttingen, Department of Agricultural Economics and Rural Development, Platz der Göttinger Sieben 5, 37073 Göttingen, Germany
raab@drv.raiffeisen.de

For reasons of food safety, there is a continuously increasing demand for innovative solutions to design risk management and risk communication. Above all for early detection of risks for the consumer, databases and communication systems have been established at national, EU and global level (Magnusson *et al.*, 2001). For example, the Rapid Alert System for Food and Feed (RASFF) was implemented to enable rapid communication between public coordinators.

Precautionary product warning shouts and targeted product recalls owing to health risks for the consumer or complaints due to insufficient quality are in the food producing chain still often occurring events necessitating intercompany risk management. In addition to health risks for the end consumer, these occurrences can result in damage of reputation, loss of market shares, turnover loss, cost increases, claims for compensation and liabilities for the companies involved in all stages of the food chain (Pfeifer, 2001; Schmitz, 2006). In addition to this, it has to be taken into account that companies in the agri-food industry are closely interlinked in value chain characterised by intensive division of labour. Control of certain risks, for example of animal disease risks, is therefore quite often not possible for individual companies on their own. In this case, it is rather necessary to make use of a concerted action in which several companies along the value chain are participating. An example of this is *Salmonella* monitoring in the meat industry. This necessitates close cooperation of agricultural enterprises whose animal stocks are responsible for transferring pathogens to the slaughterhouses, slaughtering companies usually determining the *Salmonella* status of their production enterprises, and veterinaries looking after the stock and able to support stock sanitation in the agricultural enterprises.

In recent years, it was possible to gain experience explaining how to establish PPP risk management systems against this background (see Chapter 4.1). DIN ISO 31000 can be used as a guideline for a risk management system for individual enterprises and also across enterprises.

DIN ISO 31000 defines the term risk as follows: 'a risk can be defined as the description of an event possibly having negative consequences'. Another definition unquestionably relevant for the food producing chain is summarised by Jantke (2006): 'the risk reflects the extent of negatively missed objectives of a criterion or characteristic (e.g. product safety) in the production process when a hazard occurs.' In order to systemise the risks with which the companies of the agri-food industry are confronted, the following risk categories have been developed in recent years (Frentrup *et al.*, 2010; Lehrner, 2002):

- product risks (e.g. insufficient availability of raw materials, quality problems, occurrence of animal diseases);
- market and price risks (price volatility, appearance of new competitors, loss of important customers, etc.);
- political risks (tightening of liability, animal protection or building law, negative developments in the tax or social insurance law, etc.);
- financial risks (in particular liquidity shortages, financing difficulties, interest change risk);
- personal risks (for example the absence of important employees caused by sickness or accidents);
- plant risk (e.g. fire, burglary, IT problems);
- other risks (liability claims, threatening litigations, theft, etc.).

The occurrence of risks can lead to considerable economic disadvantages for companies and possibly also endanger their existence. Therefore, systematic risk management aiming at identifying, quantifying, controlling and monitoring risks of loss is necessary. The main objective of corporate risk management is safeguarding the existence of the company. In addition to this, various secondary objectives can be pursued, such as avoiding financial risks. Quite often, a difference is made between special and general risk management. Special risk management deals almost exclusively with basically insurable risks, e.g. theft, fire or liability claims. General risk management addresses in contrast to this also all non-insurable hazards possibly affecting an enterprise, such as political risks or the occurrence of market and price risks. Risk management is a systematic procedure which can be applied for controlling all risks in a company (DIN ISO 31000, 2009). It is according to DIN ISO 31000 a task to be carried out by the management and contributes to increased performance and improved efficiency of an organisation. Thus it is possible to meet safety needs and assure that organisations and systems reach their goals. A systematic risk management process in companies encompasses four steps, i.e. risk identification, assessment, management and control (see Chapter 4.3, Figure 4.10). Within the meaning of the EU food law, risk analysis can be directly associated with risk management from the point of view of companies.

Risk identification and risk assessment

Within the framework of risk identification, potential faults and their causes and resulting damages are determined and systematically outlined in the risk description. For qualitative risk assessment, the probability of occurrence of various risks and the extent of damage to be expected are categorised in different damage classes and reflected in a risk matrix (or a risk portfolio) (Figure 8.12) (Schmitz, 2006). By means of a risk portfolio, quantitative

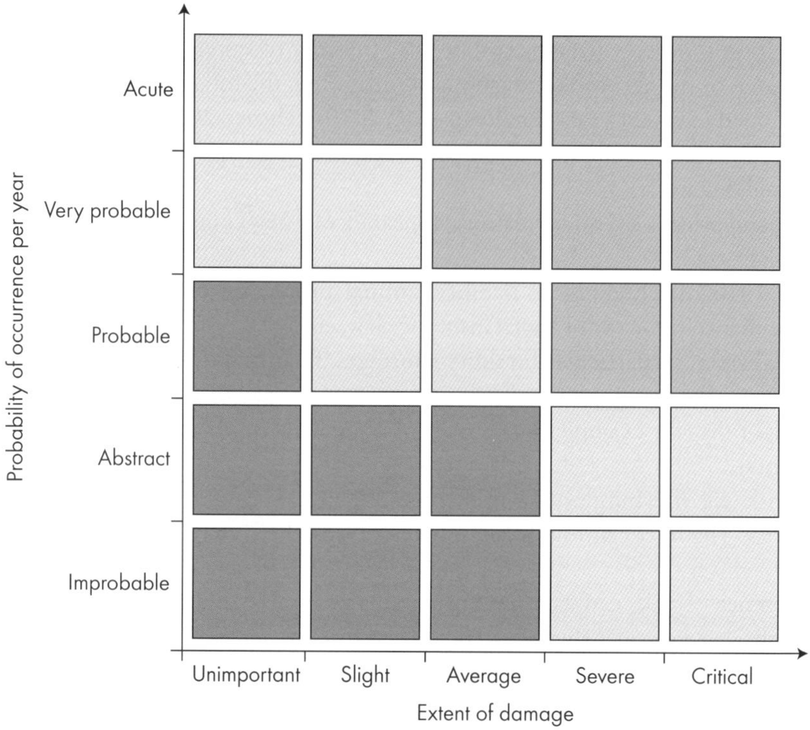

Figure 8.12. Example of a qualitative risk portfolio (Göser, 2010).

or qualitative dimensions of risks in a certain area, field of business or in a company are portrayed. The different degrees of probability of occurrence and extent of damage of a risk are illustrated by way of example as a table with 25 fields (Vanini, 2012).

Subsequently, the example of a dairy farm illustrates practical application of the described procedure within the scope of risk identification and assessment. The farm which followed in recent years a distinct growth strategy subject to certain risks has resulted from the fusion of several previously independent milk producers. In collaboration with the operation manager, consultant and scientist, the risks particularly important for the farm have been identified, assessed and illustrated in a risk matrix to verify the risk situation. The various symbols in Figure 8.13 characterise categorisation of the risks observed in the risk matrix. Some risks, for example volatile milk prices and liquidity difficulties, are distinguished by a very high probability of occurrence. Other risks, such as occurrence of an animal disease or absence of the owner-manager of the farm, have to be especially taken into account within the framework of corporate risk management since a high potential extent of damage might be expected from them.

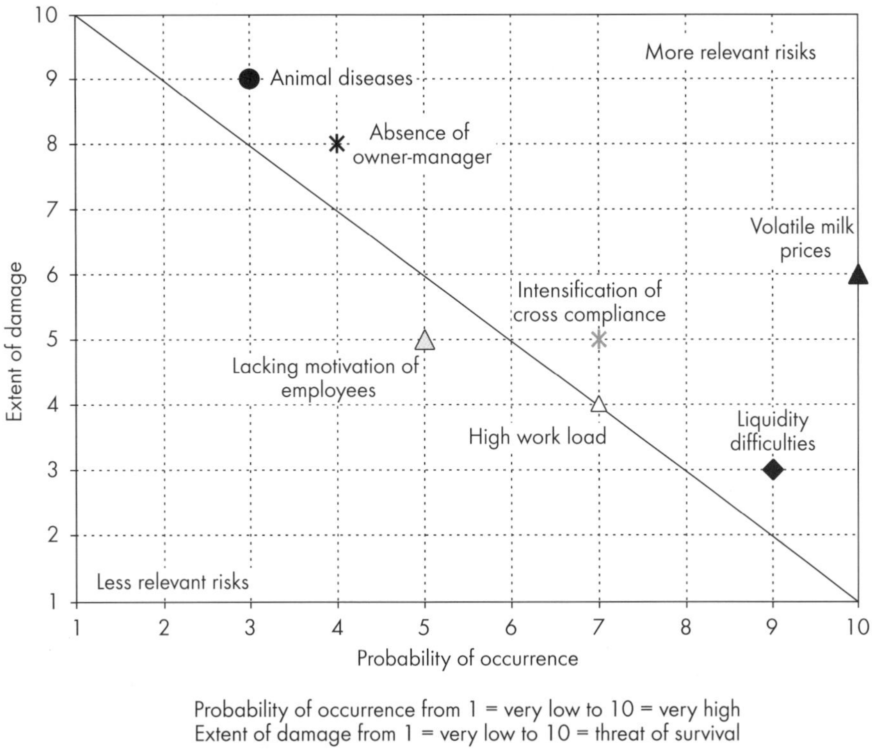

Figure 8.13. Risk matrix by way of example of a dairy farm (Göser, 2010).

A risk matrix is an instrument of risk management which can be applied to all value stages of the food producing chain and for many different threats (see in exemplary form the application in an enterprise of the meat industry with (Frentrup and Festag, 2012). Its simplicity facilitates implementation within the scope of corporate risk management and assures acceptance by the management. At the same time, it has become evident repeatedly that the risk matrix is suitable despite its simplicity to successfully identify the need for action in corporate risk management. For example, in connection with the dairy farmer mentioned above (Figure 8.13), gaps in the insurance protection for the outbreak of an animal disease could be determined.

Risk management

For the risks categorised as inacceptable within the scope of risk assessment, four alternatives for action, i.e. risk avoidance, risk reduction, risk transfer to third parties and bearing the risk independently, are available (Schmitz, 2006). The selection from among alternative risk management instruments is done under the aspect of their effectiveness, i.e. their ability to reduce a certain risk, and their efficiency, i.e. their cost in relation to their contribution to the reduction of the probability of occurrence or the potential extent of damage of a risk.

The measures taken to possibly reduce, transfer and avoid risks have to be supplemented by systematic liquidity management contrasting the deposits and withdrawals of a company oriented towards the future and thus helping to guarantee the solvency at any time – even if a risk occurs. In Table 8.7, the alternative risk measures for companies of the food producing chains are summarised.

Risk control

Risk control involves planning, control, coordination and continuous forwarding of information to the risk management. Within the scope of risk control it has to be verified, amongst others, if the expected risk reduction has been achieved by the use of an instrument,

Table 8.7. Alternatives for action during risk control with examples for possible risk measures (Pfeifer, 2001; Woidich, 2004).

Action alternatives during risk control	Examples for risk measures
Risk avoidance	• substitute product (e.g. raw materials) • retreat from sales market • using different technologies • selection of another recipe
Risk reduction	To reduce the probability of damage occurring and the extent of damage: • technological measures • crisis management, emergency planning • organisational measures
Risk transfer	Insurance: • employer's liability insurance • product liability insurance • recall cost insurance Contract layout: • general terms and conditions • quality management agreements
Independently bearing the risk	• accepted risks • non-insurable risks • non-insured risks • deductibles

at which cost this occurred and if modified framework conditions, e.g. the occurrence of animal diseases or changed competitive conditions, require adapting risk management. If the intended risk reduction was not achieved or only at unreasonably high cost, the phases of risk identification, assessment and management have to be completed once again.

Use of preventive QM methods in risk management

During practical implementation of risk management, it is often ignored that a risk management system within an integrated management system should map and assess all risks (Schmitz, 2006). Organisations of the food producing chain often apply risk management mainly as a financial controlling instrument and make no reference to other product liability cases based on other problems.

The approach of Schmitz (2006) illustrates how preventive and risk-oriented QM methods within the framework of risk management can be used. On the basis of a methodology guideline as basic instrument for consulting small and medium-sized enterprises, the two approaches of Hazard Analysis and Critical Control Points (HACCP) and the Failure Mode and Effects Analysis (FMEA) are linked with each other. The guideline facilitates structured development, documentation and implementation of an enterprise-specific self-control system.

The model for 'modularising the HACCP concept and FMEA and combining them in method modules' (Schmitz, 2006), is summarising the main steps of the HACCP concept and FMEA in a method-oriented model. FMEA, in turn, usually consists of five individual steps realised subsequently to test planning and flowing into the module 'monitoring'. Thus, from the combination of these two concepts, result the six method modules, i.e. structural analysis, failure analysis, risk assessment, risk minimisation, monitoring and verification (Schmitz, 2006).

In Figure 8.14 the HACCP und FMEA modules are compared and merged to become method modules. In the approach illustrated by Schmitz (2006), the two rigid method complexes HACCP and FMEA have been dissolved and flexibly arranged. The methodology guideline systemises failure assessment by identifying control points with unique designation by means of risk assessment:

- Critical Control Points (CCPs);
- Hygienic Control Points (HCPs); and
- Quality Control Points (QCPs).

The control points form the starting point for continuous improvement in an integrated hygiene and QM system. With the HACCP concept and with the FMEA concept, the first step involves delimitation and description of the object to be investigated. This can be for

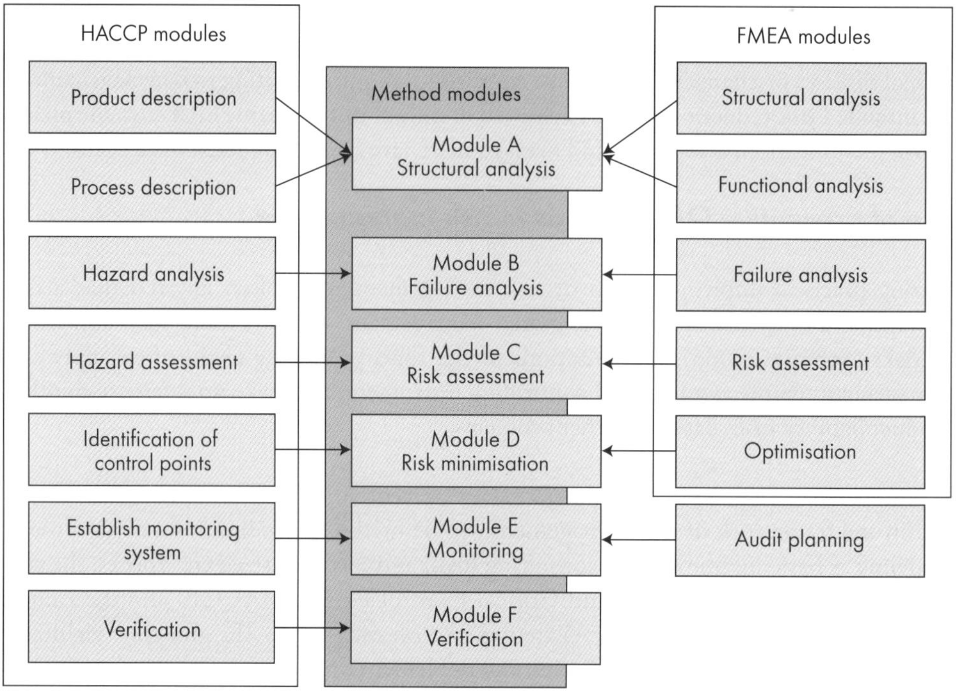

Figure 8.14. Modularisation of Hazard Analysis and Critical Control Points (HACCP) and Failure Mode and Effects Analysis (FMEA) and creation of method modules for risk assessment (Schmitz, 2006).

example a product, a product group or a process. The HACCP concept is subdivided in the approach illustrated by Schmitz (2006) in six main steps forming the HACCP modules. The seventh principle, usually designated as documentation, is not defined as an individual step here since it is accompanying all other steps. In contrast to this, FMEA consists of five main steps representing the FMEA modules. Audit planning follows as an individual module after FMEA. The HACCP modules product and process description and the FMEA modules structural and functional analysis are summarised in module A Structural analysis. Module B Failure analysis registers failure and risk analysis. The term failure is defined as 'non-fulfilment of a requirement' (Leonard and Naumann, 2002) and can be taken as the generic term for all types of failures. A hazard is within the meaning of the HACCP concept 'an agent or factor of biological, chemical or physical nature with the capacity to cause damage to health' (Codex Alimentarius Commission, 2003). The production of safe foodstuff is a requirement made by the HACCP concept. If this requirement is not fulfilled, a hazard results which is by definition a failure. The 'module C Risk assessment' contains an evaluation of the rating values A (probability of occurrence), B (significance of consequences) and E (probability of discovery) and calculation of the risk priority figure and the risk figure A × B. In 'module D Risk minimisation', the measures contributing to the control of relevant critical failures are determined and planned. The 'module E Monitoring' responds to the requirement of the HACCP concept to define for each identifiable critical control point control measures, suitable limit values, monitoring procedures and corrective measures and document them in an audit plan. This module is also applied, however, to plan the audit of quality control points (QCPs). Finally, the 'module F Verification' serves to systematically check the efficiency of the self-control system introduced.

Another team-oriented method to minimise failures within the framework of process improvement is the Six-Sigma approach. Experience shows that with consequent application of this method, cost can be dramatically reduced and turnover increased. Six Sigma (Magnusson *et al.*, 2001) was first used in the late 1980s successfully by Motorola and has since then been used by more and more companies across the industry. Many of these companies report similarly good results (Folaron, 2003; Kamiske and Brauer, 2008; Wappis and Jung, 2008). Modifying the 'Plan-Do-Check-Act Cycle' (PDCA) (Chapter 4.2), the five steps 'Define, Measure, Analyse, Improve, Control' (DMAIC approach) were defined as a model for systematic fault minimisation (Kamiske and Brauer, 2008). In each individual step, methods and tools of quality management are systematically applied. Figure 8.15 describes the five steps and provides an overview of the applicable tools (Wappis and Jung, 2008).

Klauke and Brinkmann (2009) show by way of example from the value chain pork that the statistical quality objective Six Sigma is to be applied in the agri-food industry as well. To test the use of preventive tools of the Six Sigma method along the pork producing chains, data from 190,000 pigs destined for slaughtering was used. The weights and lean meat percentages of slaughtering pigs were analysed to assess process capability and subsequent development of measures to minimise faults (Figure 8.16). In connection with further information, such as the age of the animals, information about antibiotics used, organ findings and/or results of *Salmonella* monitoring, preventive QM tools can also be used within the scope of Six Sigma for risk minimisation (Klauke and Brinkmann, 2009.

Ellebrecht (2012) illustrates one possibility for combining the advantages of the three team-oriented QM methods illustrated – HACCP, FMEA and Six Sigma approach. For this purpose, she stripped down the methods mentioned to their relevant partial steps and fit them together again in task blocks with a high degree of similarity or with specific shared tasks. She tested this procedure of combining methods using the extension of service portfolios of network coordinators in pork producing chains as an example. During a preparatory phase and based on her approach, first of all the responsibilities of farmers, veterinaries and consultants were defined as a team-building measure. In the subsequent step 2, the service processes, the definition of target values, the definition of risks, deviations and causal relationships were described and finally critical control points defined. In the

Phase	Target	Main tasks	Tools	Results
Define	Define improvement project	• Describe starting situation • Create process overview • Determine customers and requirements • Define project	• Supplier inputs process outputs customer (SIPOC) model • Project plan	• Overview of situation • Defined customer requirements • Project task
Measure	Determine actual situation	• Detail processes • Implement existing data • Register and evaluate data • Ensure suitability of audit system • Determine process performance	• Measurement system capability analysis • Pareto analysis • Process capability investigation	• Understanding the situation based on facts
Analyse	Identify relevant causes	• Determine possible main influential factors • Identify relationship between cause and effect • Identify and illustrate relationships	• Cause-and-effect analysis • Correlation and regression • Design of experiments	• Proven relationships between cause and effect
Improve	Develop and test solutions	• Develop solution variants • Assess and select solutions • Test solutions • Plan implementation	• Robust design • Creativity technology • FMEA	• Optimises tested solution • Release for implementation
Control	Implement and secure optimum solution	• Embed solution organisationally • Secure improvement sustainably • Conclude project	• Process procedure • Process control chart • Training	• Improved condition • Assessed improvement and savings • Project finalisation

Figure 8.15. Steps of the Define-Measure-Analyse-Improve-Control (DMAIC) cycle and applicable tools (Wappis and Jung, 2008).

(final) step 3, the design of measure management was determined together with tasks and time schedules.

The illustrated concepts and tools for intercompany risk management have been discussed for some time using the term of supply chain risk management. 'Supply chain risk management is to be understood as a partial area of supply chain management including all strategies and measures, information, institutions, processes and technologies suitable at a technical, personal and organisational level to identify and assess risks within the supply chain and to support conscious decision-taking' (Kersten *et al.*, 2008). In particular supply and demand risks in addition to general environmental and company-specific risks are of central concern here (Christopher and Peck, 2004).

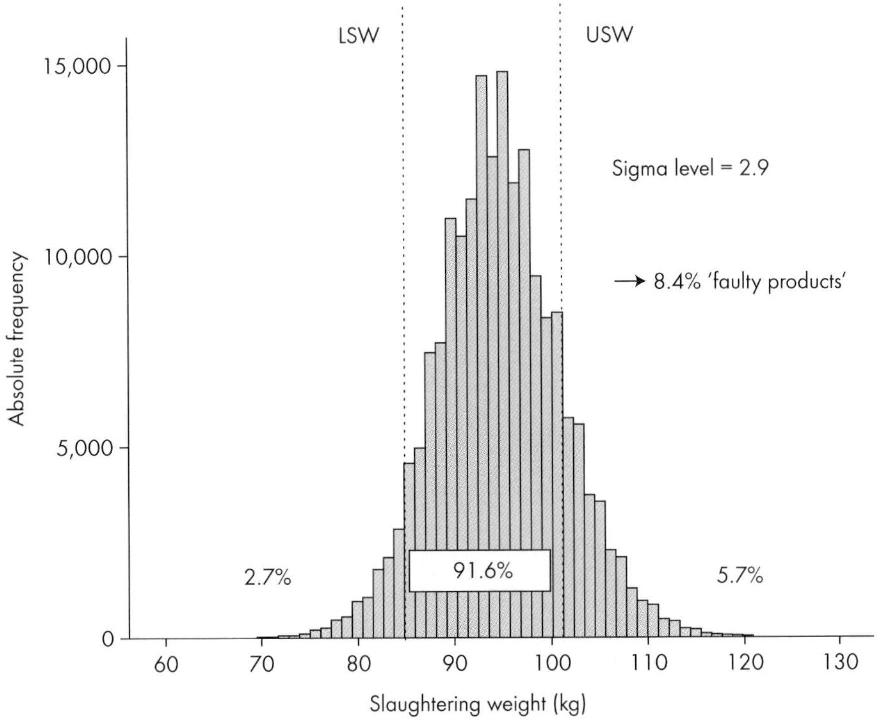

Figure 8.16. Distribution of slaughtering weight of slaughtering carcasses (Klauke and Brinkmann, 2009).

In the agri-food industry further considerations are important apart from the supply risks in intercompany risk management because for many foodstuffs the following applies: the quality of the end product already begins with raw-material production. In the milling sector, for example, close cooperation between milling companies and agricultural enterprises enables quality assurance right from the very start (Frentrup and Festag, 2012). This applies specifically to the meat industry owing to the risk of spreading animal diseases. Here it has proven very effective to systematically register risks along the value chain within the scope of a risk inventory, to assess them in a risk matrix regarding the probability of occurrence and possible extent of damage and to subsequently identify strategic starting points for the management of the various risks and implement corresponding measures (Frentrup and Festag, 2012). Table 8.8 shows the result of an appropriate risk management process by way of example of a medium-sized company operating in the meat industry.

Increasing internationalisation of value chain is also in this respect seen as an important risk driver promoting interruption of supply chains (Kersten *et al.*, 2008). Supply chain risk management is carried out analogously to the internal risk management process. However,

Table 8.8. Strategies in a company operating in the meat industry to limit risks during the production process (Frentrup and Festag, 2012).

Strategic starting point	Measure
Feedstuff program	Central purchase of feedstuff; definition of ingredients, components and recipes
Central procurement/consulting	Selection of genetics; gilt rearing enterprises and sperm; production consulting
Piglet rearing via own member enterprises	Production surplus of piglets relative to the demand of pig fatteners in the cooperative
Animal health monitoring	Regular screening for pathogens (PRRS, dysentery); *Salmonella* monitoring
Project work	Animal health management; benchmarking; animal welfare label; Isoflurane narcosis; avoiding of docking of tails; consumption of drugs
Regionality	Looking after and consulting agricultural enterprises; detailed knowledge of enterprises and procedures
Own transport	Adhering to animal protection conditions in the own sphere of influence
Respectful slaughtering	Manual production; animal protection; meat quality; social acceptance
Customer structure	Specialisation on small-trade businesses (butcheries) and quality-oriented processing companies and food retail
Tracing of origin and traceability	From closed production chain

not individual risks are assessed, but rather – with a view to supply risks – the different suppliers of a risk matrix regarding their probability of failure and the financial losses resulting from this. Risk control in intercompany risk management is targeted towards increasing insensitivity to interruptions of the supply chain, for example by cooperating with several suppliers or maintaining warehouses (Hotwagner, 2008) to ensure foodstuff chains.

Chapter 9. Recommendations on the advancement of quality management

In times of global food streams and increased vulnerability to crises, the exchange of quality-related information between the different actors of the food supply chain becomes more and more important. How can we assure that the actors communicate and cooperate intensively with each other?

This chapter will inform you about:

- how quality communication in value chains can be improved;
- what public and private organisations contribute towards crisis prevention and management;
- how quality management competences can be acquired.

Keywords: quality networks, quality communication, crisis management, ppp exercises and alliances, qm competences, international certificates.

Summary

Initiated by food crises, the companies of the agri-food industry have taken a multitude of initiatives to optimise quality communication in value chains, e.g. by the introduction of various traceability systems. The focus is in particular on transparency requested by consumers and the real use of the information provided in the value chain. Therefore, in future, further development concepts for agile and dynamic quality management networks supporting an intensive exchange of information and intercompany and inter-sector cooperation are in demand. Integration of issues such as quality, environment, safety at work, risk and social responsibility, is the challenge of the future when designing integrated management systems – not only for individual companies but above all for globally-oriented value chains. In this respect, qualification and competences of employees are important factors for the process of continuous improvement. Only in this way, the actors of the agri-food industry can successfully cope with constantly increasing requirements in global markets.

DOI 10.3920/978-90-8686-789-9_9, © Wageningen Academic Publishers 2014

9.1 Continuous improvement of communication in quality networks

L. Theuvsen[1], V. Raab[2] and S. Lehnert[3]
[1]Georg-August-University Göttingen, Department of Agricultural Economics and Rural Development, Platz der Göttinger Sieben 5, 37073 Göttingen, Germany
[2]DRV, Deutscher Raiffeisenverband e.V., Pariser Platz 3, 10117 Berlin, Germany
[3]University of Bonn, Katzenburgweg 7-9, 53115 Bonn, Germany
ltheuvs@gwdg.de

Responsibility for safe food and feed and for the quality of their production predominantly falls with industry (J. Hanf, personal communication). The higher the degree of internationalisation of goods streams and specialisation of companies within the value chain, the more elaborate and complex is communication between all kinds of customers and suppliers (Gampl, 2006; Bijman *et al.*, 2006). Therefore advancement concepts for agile and dynamic quality management networks supporting intensive exchange of information and intercompany and inter-sectoral cooperation are in demand in future. How this can be done is explained subsequently.

In every collaboratively organised value chain, there are goods, money and information streams. The information exchanged along the chain refers to various facts, amongst others also to quality aspects. Quality communication takes place between companies (business-to-business, B2B) and as well between companies and end users (business-to-consumer, B2C). The increased importance of quality-related information exchange in the agri-food industry is explained by increased requirements on safety and quality of food but also by the demands for more transparency of the production chain. The request for more transparency, in turn, is closely related to the increasing level of mistrust of the consumers towards food production owing to several severe food crises. Companies, on the other hand, recognise open communication with the end consumer as an opportunity to improve their battered image and to preserve the legitimacy of their activities in the eyes of the broad public.

Transparency describes 'the extent to which all the Netchain's stakeholders have a shared understanding of and access to the product-related information that they request without loss, noise, delay and distortion' (Hofstede, 2003a: 18). In recent years, a multitude of initiatives have been undertaken to increase transparency of the value chains in the agri-food industry. Examples for this are (Deimel *et al.*, 2009):

- The EU and also national legislation have enacted various legal regulations to improve transparency of value chains in the agri-food industry. The basis for this is the EU Basic Regulation 178/2002 (EU, 2002) which describes in Article 18 the obligation of traceability of food and feed in the meaning of the 'one step up, one step down concept' (Hollmann-Hespos, 2008) (see Chapter 8.2). Further examples are the beef labelling

law including the underlying legal instrument of the European Community and the European Union, the so-called EU hygiene package covering the regulations on food chain information (EU, 2004a,b,c) and extensive regulations on labelling of products (e.g. allergen labelling according to regulation (EU) No. 1169/2011 (EU, 2011)).
- Certification systems serving to differentiate the products on offer, inform the end user about the type of quality properties assured by the relevant label (e.g. higher animal welfare standard, fair trade, ecological production) (see Chapter 5.2). They are thus a central element of quality communication in the business-to-consumer area. Within a limited extent, they also contribute to improvement of quality communication between companies by establishing documentary obligations.
- IT systems improve for the companies and also for the end consumer access to information and hence encourage transparency of value chains. An example for this development are in the business-to-business area various information systems used by leading companies in the meat industry or offered by external software providers to improve the exchange of information between farmers and slaughterhouses (e.g. Chainfood, Qualitype, FarmingNet, Farmer's Friend, mais Infosystem Fleisch, Schlachtdaten Online). Regarding the end consumer – in addition to traceability systems (Ellebrecht, 2008; Poignee *et al.*, 2002; Schultze Althoff, 2006) – mainly the different ways of using mobile communication have become more and more important, especially when making quality-related information available at the point of sale. Under the motto 'extended packaging', for example, the possibility is discussed to forward product information by mobile phones. For this, the user scans a barcode with his mobile and is then automatically forwarded via his internet connection to the webpage of the manufacturer where he can find further information about the product (GSI Germany, 2012).

In the previous chapters, quality strategies have been presented which help to increasingly shift QM measures – in terms of preventive quality assurance – to the planning area for product and process preparation. For this, decision-relevant information and continuous communication between the actors of the value chain are indispensable. Owing to this, procedural models have been developed and tested in scientifically supported joint projects which are now transferred gradually into practice or are already used.

Especially innovations to design quality communication in value chains have been explained by means of several case studies. Continuous improvement of quality communication will focus in future more and more on strengthening the relationship between customers and suppliers, quickly solving crises jointly and withstanding the global competitive pressure.

In the following, six approaches for continuously improving quality communication in the B2B and B2C context are introduced. With the help of a numbering from 1 to 6, Figure 9.1

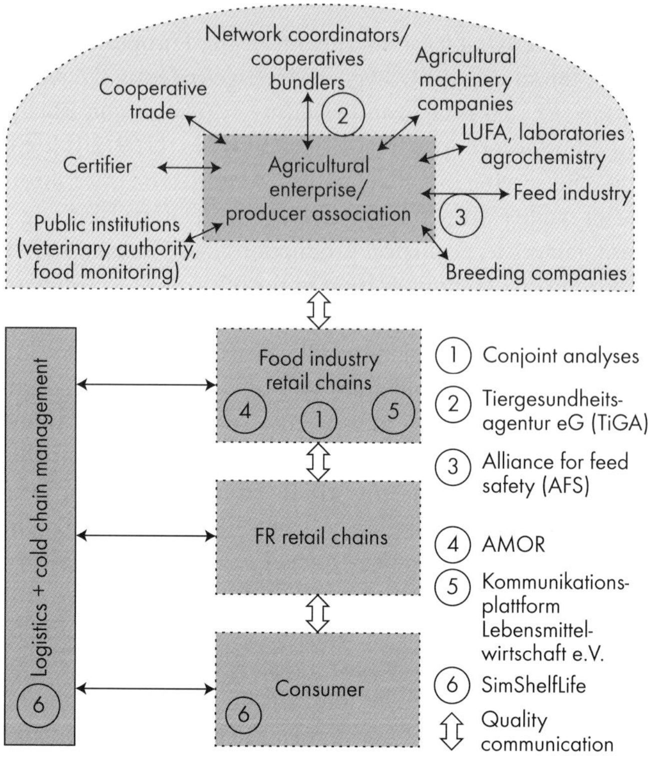

Figure 9.1. Communication in quality networks.

assigns the examples to the relevant stages or company groups actively supporting practical implementation of future concepts for quality communication.

Increasing transparency and improving access to information

Against the background of new technical possibilities and the increased range of quality information in value chains, two questions gain in importance:

- Which information and thus which degree of transparency is demanded by consumers during their purchases?
- On what depends the real use of information provided in a value chain?

The first question, i.e. how much transparency the consumers want to have at the point of sale, has been investigated by Arens *et al.* (2010, 2011) by means of an adaptive conjoint analysis. A conjoint analysis is a procedure used to survey preferences of customers or consumers (Green and Srinivasan, 1978; Wittink, 1994). Accordingly, in August 2010, 849 consumers older than 18 years were questioned via an online questionnaire regarding their perception of transparency of value chains in food production. Packaged barbecue pork meat from the self-service counter was selected as an example product. In addition to this, the statements of the test persons on subjects such as uncertainty about buying food, risk assessment, quality, perception of price and also environment and ethics were surveyed. After eliminating outliers, 737 cases were included for further evaluation. Altogether, in addition to the price, eight quality characteristics were considered in the investigation: feeding of animals, rearing, origin, information on slaughterhouse, processors and food safety (with/without certification mark), environmental standards during production and brand. Each property possibly relevant from the point of view of the consumer, was included in the investigation with different types of characteristic features (e.g. slaughtering of animals in the own region, in Germany, in Europe or no information on the place of slaughtering). The adaptive conjoint analysis subsequently enabled to present products to the test persons via computer characterised by different combinations of characteristic features. The participants in this investigation then had to evaluate the different characteristic features and thus differing degrees of transparency regarding quality-relevant product properties in order to calculate from these the individual overall benefit of each individual stimulus (Green *et al.*, 1991).

Within the scope of evaluation of the survey results, the importance of the included product characteristics from the difference between the best and worst rated typical features of each characteristic was calculated. The maximum benefit spectrum was related to the sum of the maximum benefit spectra of all nine characteristics. This resulted in the relative importance of the various product characteristics and the contributions of the various characteristics to the consumer benefit (Ahlert and Schulze-Bentrup, 2010). Figure 9.2 illustrates that the characteristics under consideration were all of similar importance for the participants of the survey. Regarding the characteristic 'price', a high standard deviation can be seen. This implies that the participants' willingness to pay more for quality-related information and thus more transparency is very diverse.

In an adaptive conjoint analysis, the respondents were also shown full profiles of products for which they were to indicate the probability of purchase. In this way, it was possible, amongst others, to verify the surveyed preference information for consistency. This phase was concluded with a calculation of the so-called individual 'calibrated utilities', the average values of which are illustrated in Table 9.1.

In the course of this, it was shown that the characteristic 'brand' was perceived altogether as most unimportant. The test persons mostly did not differ between a brand meat program ('Hofglück'), a manufacturer brand ('Kornmeyer Fleischwaren') and a retailer-owned ('gut & günstig'). Only products without any information on the brand were predominantly rated poorly. Furthermore, it becomes apparent that missing statements regarding the production process are rated negatively by all participants. This suggests that relevant

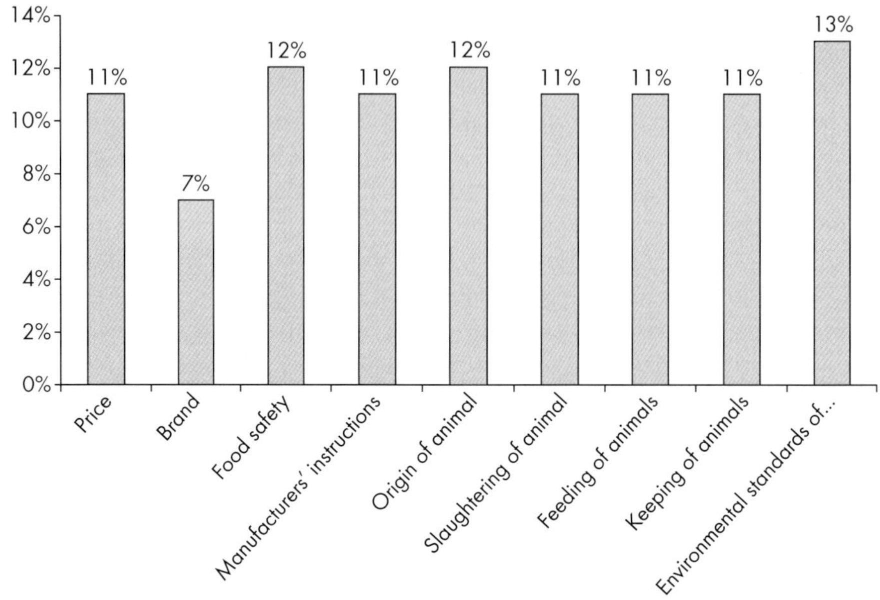

Figure 9.2. Relative importance of product characteristics (Arens *et al.*, 2011).

quality information obviously brings advantages to the consumers. This even applies to information which is more or less of no informational value such as 'from controlled production' or 'strictly adhering to legal provisions'.

The values shown in the table are partial benefit values standardised to zero. The higher the value (positive), the greater the benefit perceived. The partial benefit values of the typical features of a characteristic, e.g. brand, are interrelated. The negative sign is only relevant insofar how the partial benefit values differ from each other (difference). If the differences are as small as is the case with the characteristic 'brand', this indicates in addition a minor importance of the characteristic. The characteristic feature 'no brand' has the largest distance (into the negative) to the more comparable other partial benefit values and can therefore be seen as the characteristic feature generating the smallest benefit of the four mentioned.

Overall, the results show that the consumers are basically interested in quality communication. In this respect, they, however, prefer easily understandable, rather general information on quality, e.g. the information 'from controlled production' instead of extensive detailed information. The recommendation given to the companies of the agri-food industry is therefore to basically continue improving transparency of the value chain. In this respect, their solutions should concentrate on conveying such information to the consumers in the shops which they can easily understand and deepen if desired, for example by scanning a barcode via smartphone.

Table 9.1. Aggregated standardised 'calibrated utilities' of the product characteristic (Arens *et al.*, 2011).

Characteristic	Characteristic feature	Calibrated utilities
Environmental standards of production	beyond legal requirements	38.88
	strict adherence of legal environmental standards	27.13
	not specified	-66.02
Keeping of animals	reduced number of animals per barn with a straw bed	29.11
	reduced number of animals per barn	13.52
	strictly adhering to legal requirements	11.08
	not specified	-53.71
Feeding of animals	no use of genetically modified feedstuff	17.85
	only feed from the own farm or regional feedstuff	28.48
	strictly adhering to legal requirements	10.08
	not specified	-56.41
Slaughtering of animal	in the own region	43.94
	in Germany	24.19
	in Europe	-24.44
	not specified	-43.70
Origin of animal	from the own region	44.32
	from Germany	28.74
	from Europe	-22.09
	not specified	-50.97
Processor information	name of all companies involved	34.12
	name of farmer and slaughterhouse	25.86
	name of only one involved manufacturer	-11.55
	no name specified	-48.43
Food safety	QS - Quality scheme for food	27.66
	from controlled production	33.47
	no information specified	-61.13
Brand	Hofglück Markenfleisch	9.61
	Kornmeyer Fleischwaren	-1.13
	gut&günstig	4.30
	no brand	-12.78
Price	13.66 €/kg	-46.96
	9.50 €/kg	-14.13
	6.87 €/kg	18.91
	4.75 €/kg	42.18

The second question which came up at the beginning refers to the fact on what it depends whether the market partners in a value chain use the information provided. This question was at the centre of the study carried out by Arens *et al.* (2012) on the use of quality-related information in the business-to-business context. Concretely it was analysed if pig fatteners use information on animal health forwarded to them by the slaughterhouse taking on the animals. With the support of Qualität und Sicherheit GmbH (QS), and diverse bundlers in the QS certification scheme, 3,024 pig fatteners in total were questioned in writing in April and May 2008. 873 producers (approximately 29%) took part in this survey.

With the help of a factor analysis, seven factors potentially relevant for the use of quality information by the farmers were extracted from 28 individual statements (Arens *et al.*, 2012)

- extrinsic motivation, for example fear of sanctions by the slaughterhouse in case of a bad *Salmonella* status;
- relevance of animal welfare management, e.g. the perceived importance of animal health for the own competitive advantage;
- intrinsic motivation, namely pride and satisfaction caused by good animal health;
- professional expertise of the farmer in the area of animal health;
- frequency of recourse to different sources of information (recommendations given by consultants, *Salmonella* reports, etc.);
- general development status of animal health management in the agricultural enterprise;
- perceived quality of animal health information made available.

A multinomial logistical regression analysis showed that above all four variables influence how much time farmers spend on the use of animal health information provided to them:

- intrinsic motivation of farmers;
- quality of the information provided;
- frequency of recourse to different sources of information; and
- professional expertise of the farmers in the area of animal health.

Socio-demographic data (including size of agricultural enterprise) did not have any significance at all in the study presented regarding the intensity of the use of quality information, in this case of animal health information (Arens *et al.*, 2012).

The empirical findings on access to information by farmers show that it is decisive how information is provided by the value chain. Hence, the quality of information emerged as the central influential parameter for the fact whether farmers actually use the information on the health status of their animals and consequently contribute to the improvement of quality and safety of food, e.g. by improved *Salmonella* monitoring. The better for example the information provided by slaughterhouses is tailored to the information need of farmers, the more useful this information is perceived by the farmers and the better functions quality-related cooperation along the value chain.

Tiergesundheitsagentur eG (TiGA)

Marketing of breeding and slaughtering animals from local production has come under pressure against the background of increasing imports. By regular health checks and transparently demonstrating the health of the traded animals, the livestock trading organisations expect competitive advantages by establishing and continuously expanding their service offer via joint information and service agencies. These combine coordination tasks in quality, health and risk management (Schütz and Petersen, 2009).

As a result of the research cooperation, a supra-regionally functioning animal health management system with the first nationwide animal health database was created. This highly effective database offers all users the opportunity to manage their own animal health systems adhering to the TiGA standards at the same time. This organisation is an interactive registration body for the results from a national and standardised animal health monitoring initially from farms marketing piglets with 28 kg. The objective is, amongst others, to actively shape the communication structures at normative, strategic and operative level of quality management. As Figure 9.3 shows, the normative level mainly deals with communication in industry-specific institutions aiming at standardising procedures in quality and health

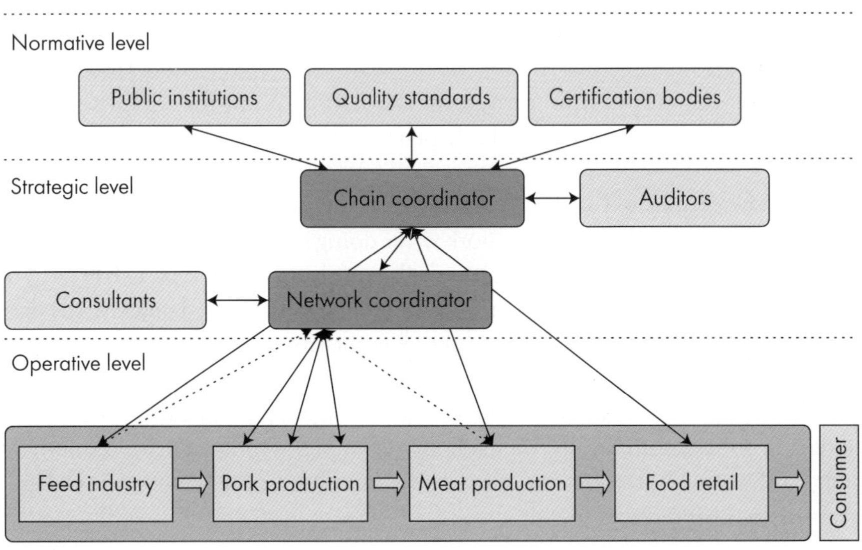

Figure 9.3. Levels of quality communication (Bahlmann and Spiller, 2009; Petersen *et al.*, 2007).

management. At the strategic level, the livestock traders see themselves increasingly in the role of network coordinators and thus as the mediator between the normative level and the companies of the value chain 'meat' at the operative level (Petersen *et al.*, 2010).

Proof of a self-control system towards the customer in the value chain and also establishing for the first time a sector-specific standard from within the production stage, are in the focus of other planned projects. For this, a uniform procedure for sampling and for the analysis of blood and faeces samples is aimed at to provide the basis for comparing providers, for example of piglet lots. This coordinated procedure is to create transparency and trust amongst German piglet producers and motivate them to participate in voluntary certification of the animal health status. In particular, the increased gain of information and the decision certainty for fattening enterprises is to be communicated as a competitive advantage. This concerted approach in connection with establishing a standardisation body is to create a signalling effect for the European countries outside Germany and especially for the European competitors.

Planning, establishing and continuously improving animal health standards follows the principles of the PDCA cycle. Each letter designates a certain phase:

- P – Plan: In the planning phase, measures for establishing and improving animal health standards are determined.
- D – Do: The planned measures are implemented in the organisations involved.
- C – Check: The measures are controlled and rated regarding their effective aims.
- A – Act: On the basis of the check result, corrective, improvement or extension measures referred to the standard are initiated, if needed.

The measures of the last phase, in turn, form the starting point for completing the cycle once again. In which way these phases are shaped by which actors is illustrated in Figure 9.4.

The bodies to be established at the normative level can be seen in Figure 9.5. The animal health standard needs not be limited to pork while doing this. For cattle whose traders are at the moment dealing with pathogens such as the bovine herpesvirus type 1 (BHV1) and the bovine viral diarrhoea virus (BVDV), such a system can be established quickly. The database entries are carried out directly by data transfer from the laboratories. These serve a so-called XML-SOAP interface. This means that data is transferred directly by the laboratories and need not be read in. This saves time and effort for the agency, at the same time verification of the data is done automatically via database (example: wrong positive results). At the moment, two major laboratories can support the interfaces; other laboratories adapt their data transfer to the interface.

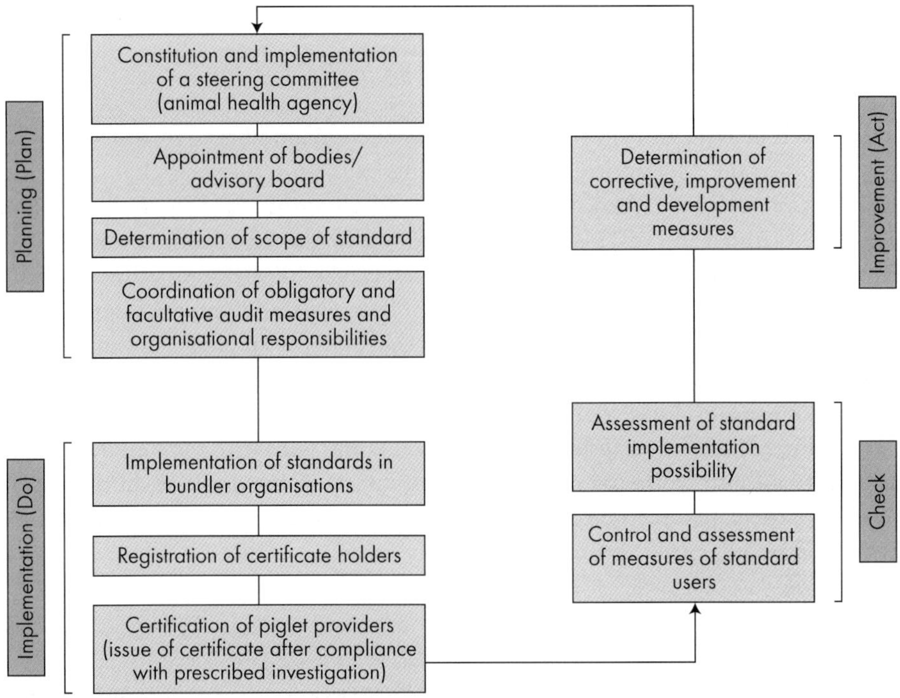

Figure 9.4. Steps of planning, implementation, control and improvement of animal health standards.

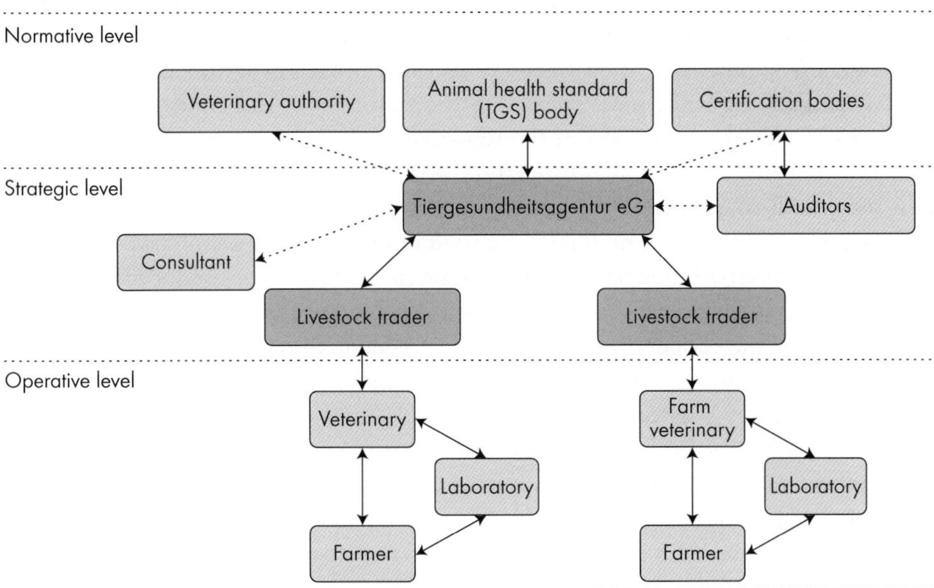

Figure 9.5. Levels of standard determination and implementation.

Allianz für Futtermittelsicherheit

17 mixed feed manufacturers with an annual production amounting to almost 2 million tons have joined the AFS initiated by a crisis situation within the value chain pork where dioxin limit values were exceeded. Within only a few months, the AFS could win almost 10% of the German mixed feed production for an organised supplier evaluation. The heart of the AFS is as well a database developed especially for the needs of the mixed feed manufacturers. At the moment, approximately 350 suppliers, 400 products and 120 users are managed via the database. In addition to this, the results of supplier audits and measures and deadlines agreed with the suppliers are documented and evaluated. The system is rounded off by the modules 'complaint' and 'damaged goods'. Complaints to suppliers are entered in the system and serve as the basis for audits and talks with the suppliers. Freights not accepted are also registered by the AFS participants in the database. In this case, the other AFS participants are immediately informed by email about bumped goods. The AFS system is thus an effective instrument of crisis prevention!

AMOR

The model AMOR (Alliances for the Mutual Organisations of Risk-oriented inspection strategies) was already explained in Chapter 4.1 as the first concept for coordinated inspection strategies defining a certain form of communication between customers and suppliers and organising a sufficient flow of information (Lang and Petersen, 2012). Although this concept was implemented initially in the meat industry, basically it can be transferred to other sectors as well. When doing so, it is important to find out how the risk awareness and the willingness to cooperate differ in various sectors. Surveys amongst business partners from the meat sector and the sectors milk and fruits/vegetables were therefore carried out with trendsetters categorised as leading regarding QM and open to new developments. 60 companies formed the preselected random sampling: 10 from the meat sector and 25 each from the sectors milk and fruits/vegetables. On the whole, 22 companies have completed the survey: 8 experts from the meat sector and 7 each from the two other sectors. Amongst the companies were cattle traders, producer associations, slaughterhouses and processing companies, private and cooperative dairies and organisations from the private and cooperative fresh fruit and vegetable wholesale.

A four field matrix – based on the BCG matrix – was developed with the two axis dimensions 'risk awareness' and 'willingness to cooperate' to compare the sectors regarding their implementation of AMOR. Both important characteristics of an AMOR inspection were described by 5 questions each (Table 9.2) which resulted in one point with a positive answer, in half a point with the answer 'partly' and in no point with a negative answer for the companies. For each axis, 5 points could thus be achieved at maximum.

Table 9.2. Elements of matrix dimensions.

Risk awareness	Willingness to cooperate
1. Are further methods apart from Hazard Analysis and Critical Control Points used for risk analysis?	Could you imagine organising an inspection jointly with other companies?
2. Do you accept external support when implementing your risk management system?	Are you already carrying out such joint inspections with other companies?
3. Do you want to improve your competitiveness with the risk management system?	Are you willing to exchange the relevant information with the supplier/ customer for carrying out an AMOR inspection?
4. Were you influenced by recent food crises such as EHEC to optimise your risk management?	Can you imagine organising an inspection jointly with the supplier mapping incoming goods inspection (at the supplier's premises) and outgoing goods inspection (within the own organisation) simultaneously to reduce the number of inspections and thus also the cost involved?
5. Can you imagine risk orientation of inspections?	Are you already carrying out an inspection which is an incoming goods inspection and an outgoing goods inspection at the same time?

Depending on the position in the matrix, the companies can be designated as 'AMOR stars' (high number of points for both axes), 'AMOR question marks' (low to medium risk awareness, high willingness to cooperate), 'AMOR freeloaders' (high risk awareness, low to medium willingness to cooperate) and 'AMOR dogs' (both areas low) (Figure 9.6).

The results of the study show: AMOR backs the formation of alliances between the actors of the chain to implement risk-oriented inspection strategies jointly. The inspection strategy contains an adaptation of the extent to the risk of the product batch for food safety. If a high risk is assumed, a more intensive inspection of the products is carried out than with a low risk. This procedure basically has four advantages since it helps

- to use scarce resources in areas with a high risk;
- to avoid double inspections of the same aspects in various stages;
- to carry out additional inspections in the chain jointly; and
- to increase the exchange of information between the individual stages.

The alliance decides jointly on the inspection strategy. This includes the organisational structure, risk-oriented design of the inspection and the distribution of tasks and responsibilities of the partners. In addition to this, an information and communication structure is developed for successfully introducing the strategy. Cost/effort and benefit are distributed to achieve a win-win situation for everybody involved. Furthermore, the

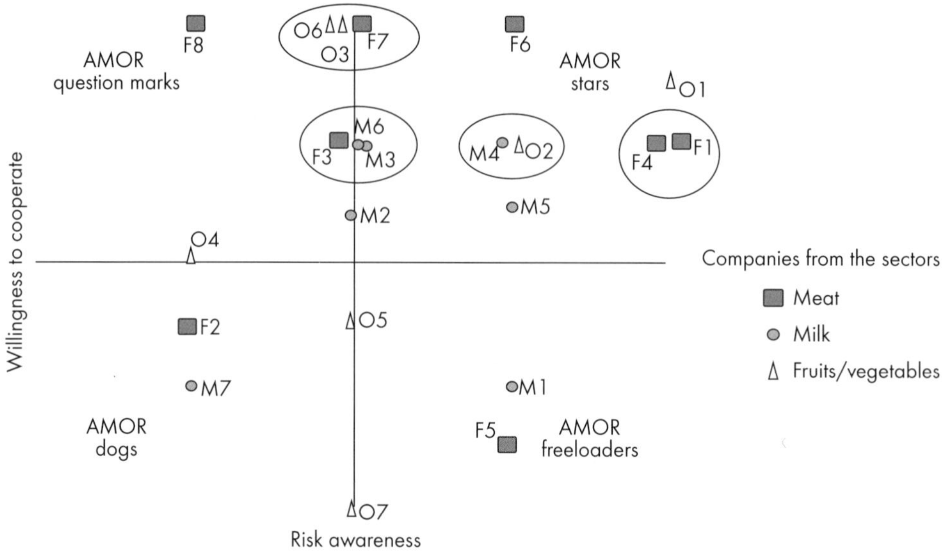

Figure 9.6. Matrix to determine Alliances for the Mutual Organisations of Risk-oriented inspection strategies (AMOR) competences.

companies are assigned to four quadrants according to their answers regarding their risk awareness and their willingness to cooperate (Figure 9.6).

The questions on risk awareness make it clear that a company has a certain way of thinking regarding the risks, and management of risks includes a lot more than just the fulfilment of minimum requirements. This is important for risk orientation of inspections. The questions regarding willingness to cooperate show the willingness to cooperate with other companies and exchange the necessary information in terms of AMOR.

You can find companies in all four fields. The largest amount can be seen in the field 'stars'. The seven companies placed there show a high willingness to cooperate with other firms regarding inspections and also have a high risk awareness. Another seven companies are positioned between the quadrants 'question marks' and 'stars'. These companies are willing to cooperate with others, while their risk awareness is at the same time only moderately pronounced. Their willingness to cooperate is decisive for AMOR. Risk awareness fluctuates noticeably and is partly dependent on acute crisis situations.

From the survey, typical encouraging and hindering factors for AMOR inspections could be deduced. It is encouraging if partner companies are part of a producer association or cooperative and long-term relationships based on trust exist between them or if they come from countries having similar laboratory and/or QM standards. In contrast to this, hindering factors for joint, risk-oriented inspection strategies are frequently changing suppliers, lack of trust in suppliers and/or customers leading to a protective behaviour, as well as a strong market and price pressure resulting in increased competition and interfering with building up trust.

The results show that the majority of the companies surveyed see AMOR audits with risk orientation and joint organisation as a realistic possibility for themselves. In addition to this, successful examples for AMOR audits can be identified. It can be determined that above all in the meat sector, strong alliances are decisive to introduce and implement risk-oriented audit strategies. AMOR reflects a new trend placing more value on transparency and quality communication. This approach forms the basis for further comprehensive research activities.

Information platform 'Die Lebensmittelwirtschaft e.V.'

Expert presentation of the production of food, on the one hand, and information on their artisanal manual and industrial processing and composition for the consumers, on the other hand, are the concerns behind the foundation of the association 'Die Lebensmittelwirtschaft' Berlin. The supporters and foundation members are the Deutsche Raiffeisenverband e.V. (DRV) and six other umbrella associations of the agri-food industry. Important improvements in food production have taken place largely unnoticed by the public. Urban lifestyle and the ever available surplus of products from around the world have resulted in Germany in the fact that basic knowledge on food which was a matter of cause for former generations has got lost. Industry has so far not been successful in conveying modern production methods to the consumer. In times of increasing mechanisation and globalisation, the agri-food industry rather acts more and more as a point of crystallisation for romantic basic attitudes. The industry must start a communication offensive to win back the ability to control its own subjects in the social discourse. Food industry is equivalent with high-tech production. It has to be consciously and offensively communicated that high tech and animal rearing appropriate to the species, environmental protection, sustainability and food safety and quality are not mutually exclusive and that modern standards are rather the necessary precondition to achieve these goals. Therefore, amongst others, the Deutsche Raiffeisenverband e.V. has supported foundation of the information platform 'Die Lebensmittelwirtschaft e.V.' which is to inform in future about all subjects along the production chain. Ignorance and fears of the public based on lacking expert knowledge in the area of modern, industrial production of food are to be reduced and the confidence of the consumers in the production and quality of food guaranteed in this way.

Web-based simulation tool to support cold-chain management (SimShelfLife)

An efficient, intercompany cold-chain management is necessary to ensure safe easily perishable food of high quality in every stage of the food producing chain (Raab *et al.*, 2008; Smith and Sparks, 2004). Since temperature is the most important influential factor regarding durability of a product, efficient temperature monitoring belongs to the most important aspects of the inspection planning in intercompany quality management (McMeekin *et al.*, 2008; Nychas *et al.*, 2008; Taoukis *et al.*, 1999; Wells and Singh, 1989). Despite various efforts, it happens in all stages of the value chain that temperature is exceeded significantly. One possibility for better control of the cold chain at individual product and batch level are time-temperature indicators (TTIs; Figure 9.7) (Kreyenschmidt *et al.*, 2003; Kreyenschmidt and Ibald, 2012). After assuring all technical, organisational and functional requirements (see Chapter 4.3, Figure 4.16) for the use of such indicators, these can be employed for support at the decision points illustrated in the process model of the exchange of goods and information between suppliers and customers (see Chapter 4.1, Figure 4.2).

Integration of mathematical models is required to calculate describing, prescribing and predicting information at the relevant decision points. Studies on the precise kinetic behaviour (Kreyenschmidt, 2003) and modelling of discoloration of the TTIs (Kreyenschmidt *et al.*, 2010) necessary for these, provide information about compliance with cold chain requirements and decrease of loss of freshness of foodstuff (Kreyenschmidt *et al.*, 2003). By reading the colour value, e.g. by means of the spectral photometer, this

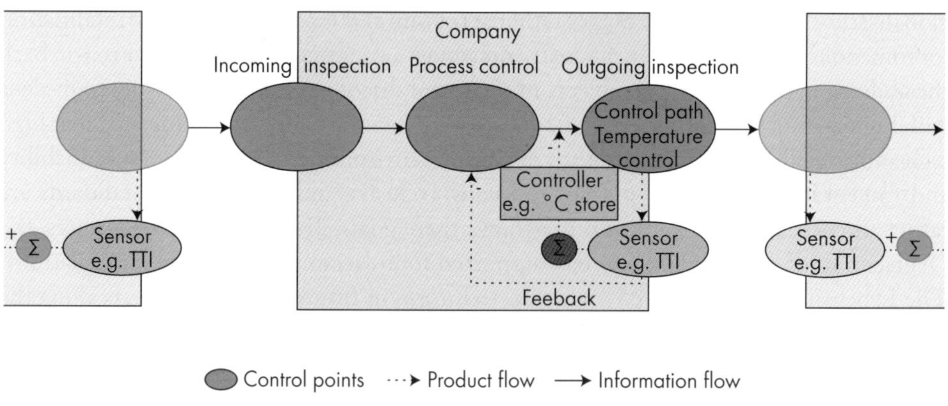

Figure 9.7. Schematic illustration of the inspection model to use time-temperature indicators (TTIs) within the framework of continuous improvement of cold-chain management (Raab, 2011 modified from Petersen, 1985).

digital information can be processed for decision finding and consulted for optimisation of further distribution, storing, etc. (Kreyenschmidt, 2003).

Use of the inspection model for cold-chain monitoring by means of TTIs has been tested in several pilot studies in the meat producing chain (Figure 9.8). The use in several pilot studies has shown that intercompany inspection models enable decision support and continuous improvement of cold-chain management by the exchange of advance information and feedback between customers and suppliers (Raab, 2011). The colour values read out in every step of the value chain by means of a spectral photometer, e.g. the indicator OnVu™ TTI, are evaluated with the mathematical model furnished and converted into decision-relevant parameters, such as decrease of the sell-by date if temperature is exceeded for a short time (Kreyenschmidt, 2003; Kreyenschmidt *et al.*, 2010; Raab, 2011).

For the actors of the cold chains, this means possibilities for optimisation of distribution and storage. Simplified processing and input into a software program is, however, necessary so that companies in various stages of the cold chain can use the model. The simulation tools necessary for this are developed by an inter-disciplinary working group 'cold-chain management' within the framework of international joint projects (Bruckner *et al.*, 2013; Eden *et al.*, 2011; Ibald *et al.*, 2011; Kreyenschmidt *et al.*, 2003; Raab *et al.*, 2008, 2011). By means of the software program, specific requirements on an efficient cold-chain management are enabled for the user:

- simulation of sell-by dates for specific products subject to dynamic time-temperature control;
- simulation of a real or assumed time-temperature relationship for planning and adapting distribution and storage;
- when using TTIs for temperature monitoring, the software enables simple monitoring if the cold chain has been respected and calculation of sell-by dates based on this;
- software is compatible with most of the commercial internet providers and can be updated and managed easily.

The relevant information is processed specifically for the individual users. A demo version can be tested at: www.ccm-network.com.

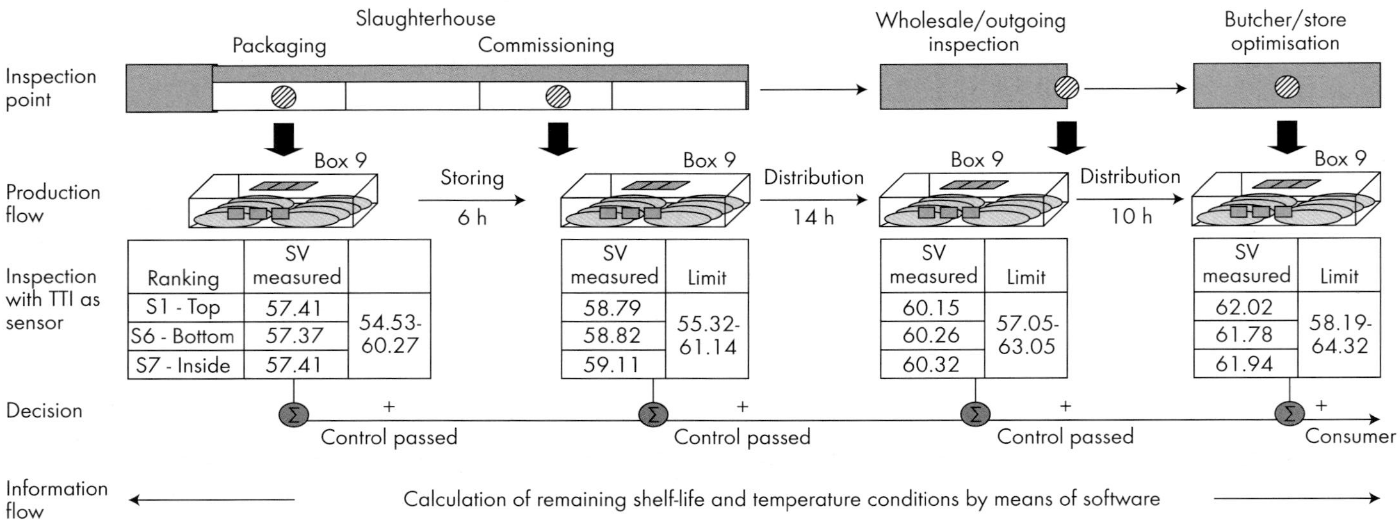

Figure 9.8. Exemplary use of the intercompany inspection model for quality communication using time-temperature indicators (TTIs) as sensors (Raab, 2011).

9.2 Advancement of approaches in crisis management

S. Slütter[1], V. Raab[2] and B. Petersen[1]
[1]*University of Bonn, Katzenburgweg 7-9, 53115 Bonn, Germany*
[2]*DRV, Deutscher Raiffeisenverband e.V., Pariser Platz 3, 10117 Berlin, Germany*
s.sluetter@uni-bonn.de

During the next 20 years, a drastic increase of global trade with feed and food is to be expected. This presents a challenge for the agri-food industry, food retail and the public sector to design supporting transnational safety systems more economically, reliably and internationally compatible. This will also include in future technical, personal and organisational development better coordinated in the form of a structural innovation for quickly solving crisis situations (Petersen, 2012b). In addition, this also encompasses development of quality management as an integration platform for further management standards at industry level. An integrated management system (IMS) provides a consistent guide frame for this, and not only for economic enterprises integrated in quality control, but also for public institutions involved in preventive consumer protection.

Integrating subjects such as quality, environment, safety at work, risk and social responsibility, represents the challenge of the future when designing the IMS – not only for individual companies but above all for globally oriented value chains (see also Chapter 2). In the past, these requirements were often considered on their own and illustrated in separate systems and structures.

In an environment becoming increasingly more complex, speed, flexibility, operative perfection and agility are expected from those responsible. This can only be assured optimally, if the links and interactions of processes are understood and the varying requirements of customers, legislation, society and intercultural life are integrated.

Figure 9.9 explains how the roles of companies and authorities and responsibilities in quality management and sanitary consumer protection will advance in future. The common objective is continuous improvement of food safety with an extension of self-auditing carried out by industry and simultaneous reduction of public areas of responsibility. After the entry into force of the General EU Food Law at the beginning of the century, this development opened up a broad area of action for public-private partnership (PPP) activities increasing continuously (Wilke *et al.*, 2013). In the following, by means of four examples possibilities for action within the PPP structures are illustrated and explained.

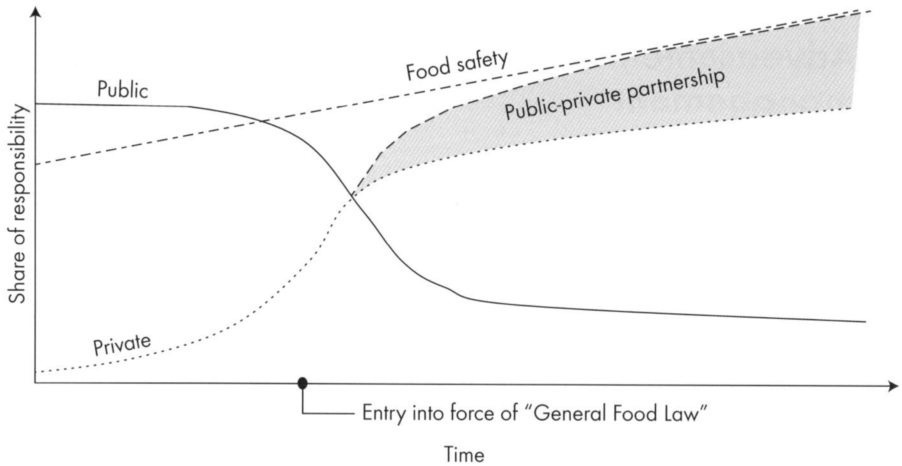

Figure 9.9. Responsibilities and chances of public-private partnerships (Wilke *et al.*, 2012a).

PPP concepts in collaborative research

The research and development platform of Grenzüberschreitende Integrierte Qualitätssicherung e.V. (GIQS) can be seen as an example for implementation of sustainable PPP structures (Figure 9.10).

Figure 9.10. Public-private partnership (PPP) research and development platform Grenzüberschreitende Integrierte Qualitätssicherung (GIQS).

When trading with animals and food of animal origin provisionally ceased completely in a difficult time of crisis for the European agri-food industry, the international association GIQS was established. Since 2001, within this cross-border research platform, 14 joint projects with almost 150 partners from industry, authorities and academia have been carried out. What originally started as several partners working together on only one project, has developed over the years into a fixed organisational structure, four long-term research alliances – Alliances for the Mutual Organisation of Risk-oriented inspection strategies (AMOR), Alliance for training and further training FoodNet Systems (ALFONS), Alliance for information and service agencies (AIDA), Alliance for engage-exchange model (A3M) – and a regional cluster on the subject of food and human safety (Bonn Research Alliance of Innovative Information Systems in International Food Chain and Crisis Communication, in short Bonn.realis) (Chapter 9.3). The latter serves, in turn, as an innovation cluster for further planned national and international networks (FoodD-A-CH and Foodbest). The task of PPP research platforms is to support their partners from industry, academia and administration to plan, prepare and coordinate development projects and process their results for different target groups in the public and private field. Meanwhile, the PPP organisation has become a constant in joint research in Europe. For this role, GIQS was awarded in 2002 the cooperation prize 'Science to Business' of North Rhine-Westphalia. Since 2008, GIQS is amongst the members of the initiative 'Competence networks Germany' and was awarded by the Federal Ministry for Economy and Technology as one of the most innovative and potential national competence networks with technological orientation. Since 2012, GIQS is amongst the initiators and founding members of Bonn. realis and FoodD-A-CH. In this function, the research and development platform supports projects for the establishment and continuous improvement of integrated management systems in value chains of the agri-food industry. This is due to the fact that GIQS excels in particular in uniting consortiums along entire national and international value chains. Continuous quality management from the field up to the counter is the precondition for healthy and safe food and the basis for effective crisis management. Technological focus of the competence network GIQS is directed mainly towards developing information and communication technologies adapted to the industry as well as modern safety technologies and methods of analysis.

Any other PPP approaches – referred to the areas of risk identification and minimisation as well as crisis prevention and crisis management – have developed from the idea of interdisciplinary research teams of the alliances AMOR, AIDA, ALFONS and A3M. The experience gained from research projects with the approaches described in the following, have motivated all participants to continue working on the concepts.

Minimisation concept as a PPP approach for risk minimisation

Pesticide residues are a subject with wide public resonance within the framework of food safety. Owing to this and initiated by a sector-wide crisis situation, large food retailers carry out since 2006 their own extensive laboratory investigations within the scope of their supplier management. This is done in addition to the monitoring activities performed by state and federal authorities yearly. Increasingly, product-related data are collected by public and private organisations and stored in their own databases, processed and analysed to assess and minimise risks occurring through the use of pesticides but also through animal drugs from primary production. Schulze Althoff *et al.* (2006) demonstrate the effectiveness of a privately organised monitoring and certification system to minimise fruit and vegetable residue contents on the basis of analysis data from a retail discounter. This basic concept may be used for other substances such as, for example, antibiotics and sulphonamides in animal production.

The authors propose a continuous minimisation process jointly supported by supply chains to the food retail and authorities for the reduction of residues in areas where so far no clear limit values of food control for approved substances in plant protection or animal health exist (Figure 9.11). This concept also makes sense in areas where limit values referred to food exist, but environmental influences and environmental effects which so far played a less

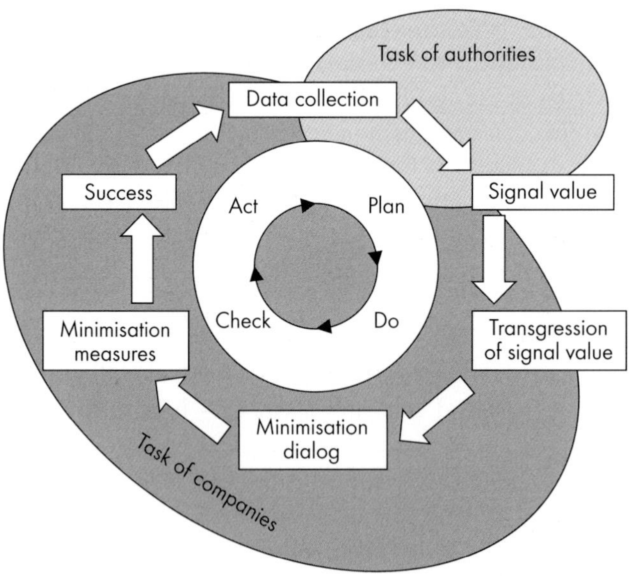

Figure 9.11. PDCA (Plan-Do-Check-Act) cycle as control cycle used in the minimisation concept (Schulze-Althoff *et al.*, 2007).

important role in connection with the approval of substances, however, have to be categorised as critical regarding their total assessment. The basis for the exchange of knowledge and data between industry and authorities is moderation of a jointly organised round table consisting of experts determining signal values based on the results of investigations from primary production, trade and food monitoring and enter into the dialog with these target values via minimisation measures. Within the framework of this PPP initiative, the actors involved try to support jointly a gradual minimisation of maximum residue limits. This PPP structure consequently serves to create responsibilities for a consumer-oriented 'risk culture' and continuous risk minimisation. If moderation is taken over by a neutral person or organisation, this encourages the process of consensus building. Additionally, a sufficiently large data basis as the foundation for targeted discussions is necessary. For this purpose, a database structure is developed which cannot only access data of business partners in the value chain in the different phases of risk management, but at the same time also specific neutral research and development databases with prognostic information from simulations (Petersen *et al.*, 2012) (Figure 9.12).

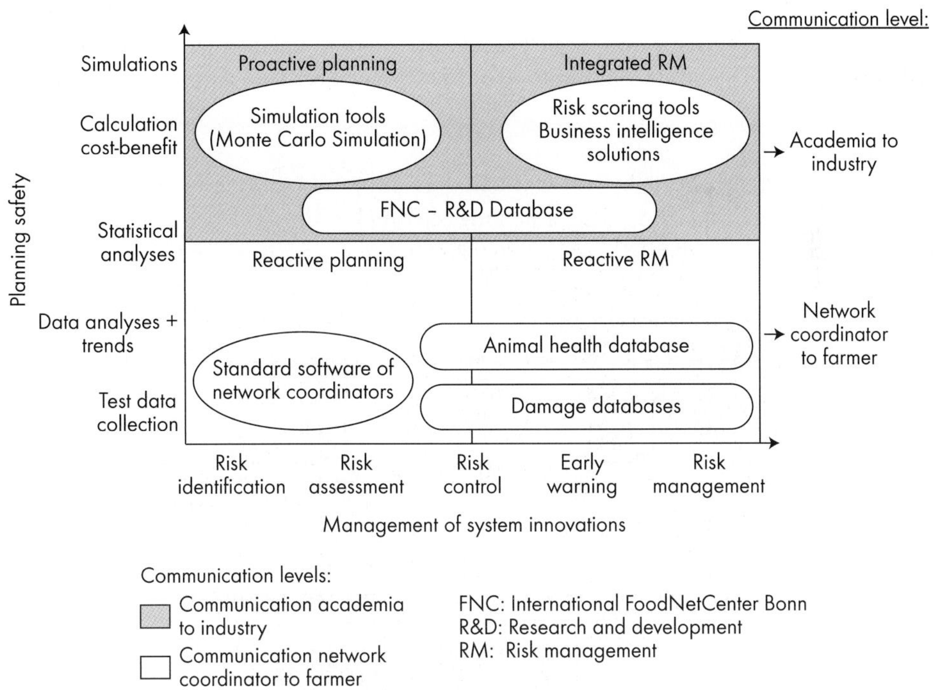

Figure 9.12. Software and hardware components to support planning safety and system innovations (Petersen *et al.*, 2012).

A strategy for crisis prevention and crisis management

Another proposal is a strategy to improve cross-border and intercompany crisis communication. The aim is to jointly design and develop technical solutions to shorten the reaction time of public and private decision decision-makers in exceptional situations (Petersen, 2012b). When establishing 'alliances to engage and exchange decision-relevant information in particular threatening situations' (3A strategy), it is a question of supporting crisis management by an improved information management (Chapter 9.3). Realisation of an electronic information exchange between industrial systems is currently established. Here, interdepartmental communication of public organisations of food and animal disease control as well as health authorities is in the foreground. For this, in PPP research projects (e.g. SafeGuard[10]) technical and organisational design proposals have been elaborated and those data identified which are required in decision situations (Figure 9.13). In addition to this, a schedule is determined to assure that time-critical decisions are no longer delayed by administrative or country frontiers or technical obstacles (Breuer *et al.*, 2008). This is exclusively data necessary for decisions in case of a crisis. It is activated for a short time just for this purpose and exchanged between the partners of the relevant alliance according to determined rules and escalation levels (Slütter, 2013). Implementation of this PPP concept requires establishment of an alliance in 'peace times' by personal contacts and confidence-building measures. Then the corresponding responsibilities are determined and guaranteed for times of crisis.

Technical realisation is done by internet-based IT support enabling interfaces between systems for data exchange and at the same time offering location-independent inquiry and entry of supplementary data. There already exist interfaces between the relevant official systems so that in connection with technical realisation for communication within the alliance only one of these systems has to be used. Particularly suited for technical realisation might be the system Balvi iP (documentation of official control) owing to the data size and the Integrated Data Processing System Consumer Protection NRW (IDV). Access to IT support in the private sector is done via interfaces with internet-based systems or direct log-in. Within the alliance, crisis exercises are organised with all involved decision-makers at regular intervals to continuously improve communication structures also in times outside crises.

Public-private partnership exercises for crisis management

Within the existing cooperation of actors from the private and public sector and from research, a common concept for cross-border or cross-country crisis exercises has been established. Here, different crisis scenarios with a reference to food safety and to sanitary

[10] http://safeguard.giqs.org.

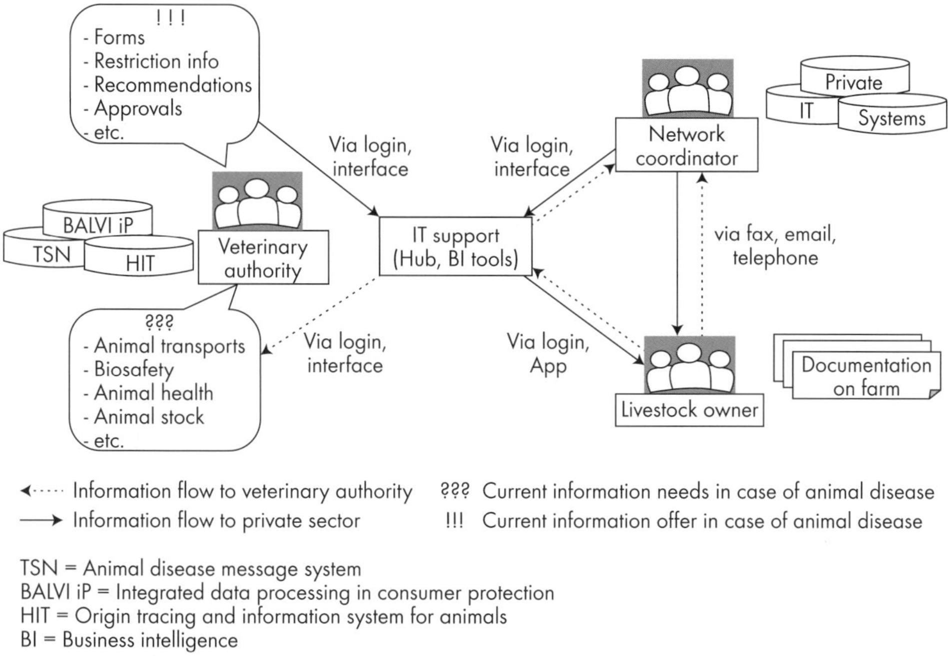

Figure 9.13. Technical realisation of communication channels within a strategic alliance for animal disease prevention and management (Slütter, 2013).

consumer and civil protection are described in detail and with possible damage situations. These are current, publicly financed projects such as SafeGuard (cross-border animal disease exercise), SiLeBAT (cross-country crisis exercise, bio-agro terrorism), Lükex 13 (food safety) where different experts from academia, administration and partly also from industry are involved. Planning, implementation, exercise framework and evaluation reports are within different responsibilities. For the future it is proposed to use synergies more frequently. In particular expert panels are to synchronise activities more efficiently to coordinate scenarios and to assess the effects for society as a whole as well as risk and crisis communication. Figure 9.14 shows how the completely different crisis committees and task forces already acting separately at various levels of the public and private sector can be networked for exercise purposes by using the existing PPP platform GIQS. Knowing who on each side is responsible and can take decisions in a crisis situation already helps to improve crisis communication in emergencies.

The objective is to develop the PPP coordination platform into an engine of strategic crisis management in Germany and Europe in this way. Coordination is done with regard to the four phases plan-do-check-act illustrated in Figure 9.15.

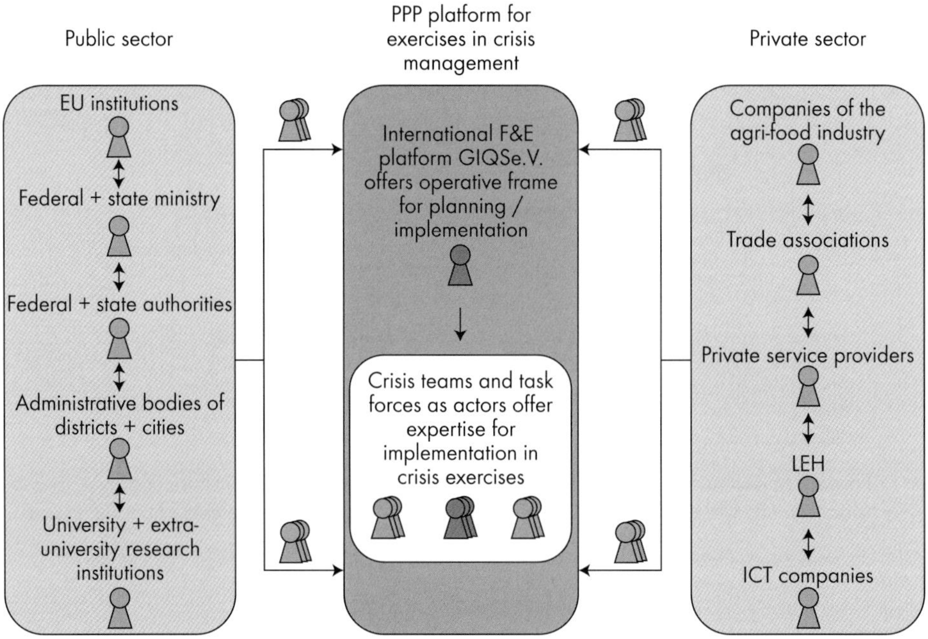

Figure 9.14. Public-private partnership (PPP) platform for cross-country and cross-border exercises in crisis management.

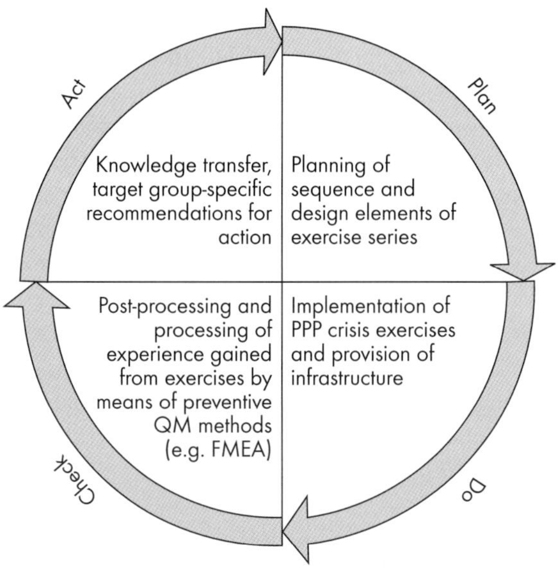

Figure 9.15. PDCA (Plan-Do-Check-Act) cycle as control cycle of Public-private partnership (PPP) crisis exercises.

In addition to this, a trustful exercise culture adapted to the new safety requirements is to develop, integrating as many decision-makers as possible from industry and administration into a corresponding training and further training program. Also, technical guidance and use instruments are to be developed in this specialised field within the framework of safety research.

9.3 Strengthening of the international dialog between industry and academia

V. Raab[1] and B. Petersen[2]
[1]DRV, Deutscher Raiffeisenverband e.V., Pariser Platz 3, 10117 Berlin, Germany
[2]University of Bonn, Katzenburgweg 7-9, 53115 Bonn, Germany
raab@drv.raiffeisen.de

For consumers throughout the world, quality and safety of food has the highest priority. In this highly sensitive buyer market, organisation and proof of inter-stage quality communication accompanying production are amongst the most important success factors to withstand international competition and to open up new markets. The complex subject is associated with the need for coordination not only between the partners of the value chain, but also with academia. This dialog is facilitated by joint research and development databases and neutral moderation by scientists (Chapter 9.2).

From the experience gained during introduction of new principles and processes in intercompany quality management, such as the QS inspection system (see Chapter 2.3 and 5.2), companies know that quite often problems arise in practice if competences and the required network structures are not planned systematically and in a structured manner. In future, strategic planning and shaping of three communication levels will be increasingly in the foreground to achieve system innovations in quality and risk management:

- communication between customers and suppliers in global value chains to strengthen international partnerships;
- communication between academia and industry to develop strategic framework conditions;
- communication between international experts to jointly manage global challenges.

In the past, it has often been neglected to prepare complex change processes affecting the entire value chain up to the food retail correspondingly and accompany them scientifically. This often resulted in instable processes – which are partly still instable, since managers, key players and above all farmers do not have sufficient time to grow into their areas of responsibility correspondingly owing to the implementation phase being too short. Effective implementation plans coordinated with academia and industry are still missing nowadays in many areas, in particular when integrated management systems are involved (Chapter 9.1).

When structured properly strategically, the development process can be subdivided into three phases from building competences through suitable co-operations up to innovation,

(Figure 9.16) and established similar to a maturity model as known from information and communication sciences:

- build-up phase;
- stabilisation phase; and
- phase of continuous improvement.

The maturity stages illustrated in Figure 9.16 can be defined with the corresponding targets to be reached within time.

For reaching the maturity stages 2 to 4, qualification of actors in the value chain, definition of responsibilities within the three levels of communication processes and finally establishment of an informal network management between the three levels of communication, time and sufficient resources are needed.

Mostly, in agri-research politics awareness has to be developed for demonstrating that the path taken to achieve system innovations offers great potential for strategic development not only of the German but also of the European agri-food industry.

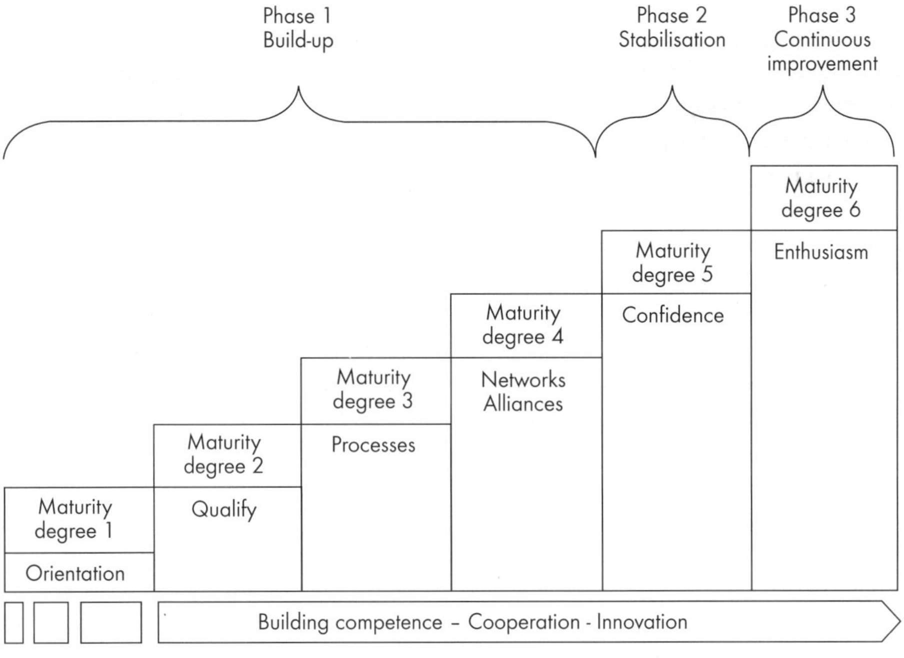

Figure 9.16. Maturity development from competence building up to innovation in quality and risk management of global value chains.

When it comes to quality communication, the willingness to cooperate is important. The future-oriented establishment of a long-term cooperation between research institutions, network managers and companies provides opportunities for strengthening the innovative power sustainably. This can assure competitiveness in the long run since globalisation of the markets and the associated competitive pressure demand from the companies but also from research institutions more flexibility and innovations. In this connection, the aspects of shortened innovation times, continuous product and process diversification, rapid technological progress, lifelong learning and changing customer requirements as well as legal regulations are amongst the framework conditions where researchers and industry experts assume responsible positions. This presents the chance to bundle the complexity of knowledge necessary for future-oriented system innovations. Cooperation is required especially where the area of competence of individual companies becomes increasingly smaller. This refers mainly to the subjects of Corporate Social Responsibility (CSR) and crisis management if responsibility is assumed for ethical and social areas implying a consensus in the entire value chain.

The small and medium-sized companies in the value chains, such as agricultural enterprises, small trade businesses and regional processing companies often do not have the money to carry out their own research. In joint projects, research alliances und public-private partnership platforms they need to be represented, however, to advance ideas. The interplay of the following is therefore important for research cooperations, i.e.

- several *users* (U) so that results can be used not only for one application but for various products and processes (broad effect);
- several *system developers* (S) delivering compatible systems consisting of hardware and software (broad safe technological base);
- several *research institutes* (R) having done basic interdisciplinary groundwork and are now participating to implement the results (state of the art);
- several *providers* (P) forwarding the results in their role as multipliers especially to agricultural and small trade companies;
- several *associations* (A) insisting on broad applicability of results (broad-scale success);
- *innovation broker* (IB) as mediators between companies and scientists;
- *public institutions* (P) at different administrative levels; and
- *network managers* (NM).

Depending on the degree of internationalisation and the complexity of public-private partnership (PPP) structures to be supported, three types of research cooperations can be distinguished:

- regional projects;
- cross-border alliances; and
- international public-private partnership networks (Figure 9.17).

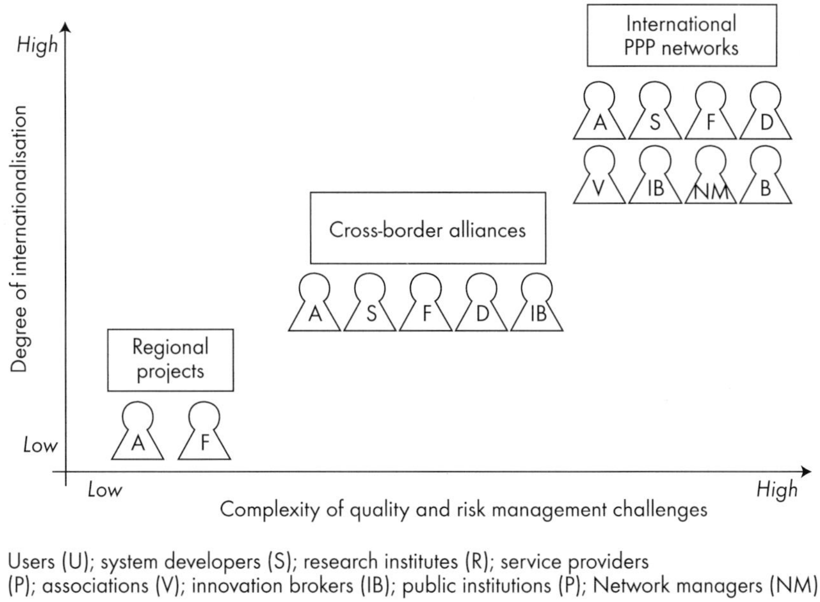

Figure 9.17. Research cooperations subject to degree of internationalisation and complexity of quality and risk management challenges.

The four private-public partnership alliances AMOR, AIDA, A3M and ALFONS are examples for cooperations initiated by an international research and development platform in quality and risk management. They were founded by a European university network and by associations of the agri-food industry (see Chapter 4.1 and 8):

- *AMOR*. Alliance for the Mutual Organisation of Risk-oriented control strategies between customers and suppliers in value chains of the agri-food industry. Here, especially companies and research institutions design intercompany inspection strategies to improve food safety and cold chain management (see Chapter 8).
- *AIDA*. Alliance for information and service agencies for horizontally bundling coordination tasks in quality, health and risk management of the meat industry. The extension of technical and organisational coordination tasks of service providers for agricultural enterprises is in the foreground here, as well as international cooperation for developing and networking of national animal health agencies (Chapter 9.2).
- *A3M*. Alliance of engage-exchange model to support crisis management and improve communication between self-control systems of the industry and public organisations of food monitoring and animal disease prevention. Here, design proposals are developed to ensure that time-critical decisions are not unnecessarily delayed by administrative limits or technical obstacles in future (Chapter 9.4).
- *ALFONS*. Alliance for training and further training to support junior staff in industry and academia in the European cross-sectional area of quality communication. In this

instance, the university network tunes its teaching offers to the new professional fields in the agri-food industry.

The basic preconditions for participating in innovation programs is the initiation of project consortiums, research alliances or clusters with the objective to bundle existing potentials from industry and academia (J. Hanf, personal communication). When the conditions are favourable, cooperation within the period of eligibility for all actors involved will be so successful that even after this support, an effective cooperation culture to the benefit of all project partners is desired.

In this respect, networks and clusters are considered as one of the possible solutions to respond most quickly and effectively to new challenges which are part of innovation processes and globalisation of the industry (Gampl, 2006).

Characteristics of clusters or networks

Clusters and alliances can be defined from an economic point of view as networks of actors along the value chains with regional concentration but with a nationwide, national or international radius of activities. The ideal and typical composition of actors is illustrated in Figure 9.18. In addition to research, training and further training institutions and service providers, entrepreneurs of one or several value chains form the basis of a cluster. Innovation-friendly framework conditions are characterised by thematic focussing of the network and transparent organisational structures with controlled management processes. Intensive communication and cooperation between the actors involved forms the basis for efficient technology transfer and thus for conversion of the investigation results into innovative products.

'Innovations move us forward! Innovations enable developments in leaps and bounds. Innovation is the older sister of continuous improvement! Continuous improvement comes about more slowly, but without improvement, innovation is a blunt weapon!' With this saying coined by Bijman *et al.* (2006), the interaction between quality management and innovation can be summarised very accurately.

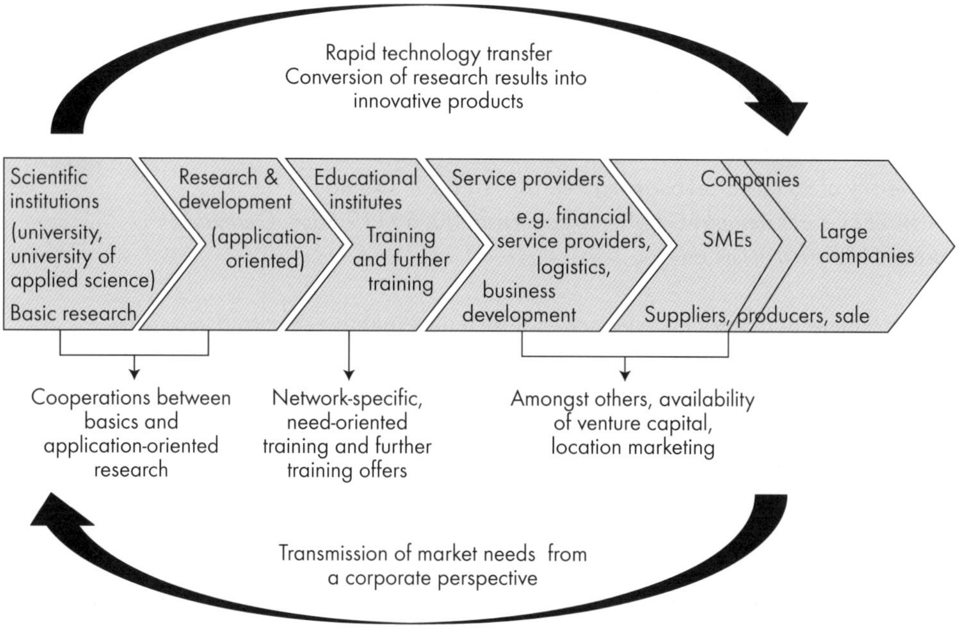

Figure 9.18. Ideal and typical composition of actors in networks (Gampl, 2006).

9.4 Advancement of quality management competences

S. Lehnert[1] and S. Bruckner[2]
[1]University of Bonn, Katzenburgweg 7-9, 53115 Bonn, Germany
[2]AGRIZERT Zertifizierungs GmbH, Vorgebirgsstr. 80, 53119 Bonn, Germany
susanne.lehnert@t-online.de

Advancement of quality management competences mainly refers to three areas:
- training and motivation of employees;
- role of quality management;
- evidence of competence by international certificates.

Training and motivation of employees

Without well trained, qualified personnel, a functioning intercompany QM system is not conceivable. However, this statement not only refers to quality representatives but to all employees in the companies of the value chain. The employees are in the centre of developments also in the agri-food industry. Employee orientation can be understood as a basic attitude according to Kamiske and Brauer (Kamiske and Brauer, 2008) 'where each and every employee is seen as an important problem-solving and creativity potential and is treated accordingly'.

The know-how of employees is thus an important factor for the process of continuous improvement. Therefore the management of a company has the responsibility to hire appropriately qualified employees and to train the staff and offer them further education opportunities. At the same time, further training is an important factor for the motivation of employees and their loyalty to their companies. In addition to this, the reasons for trainings and further trainings can be:
- differences in the actual/target comparison of qualifications;
- new objectives/improvement measures;
- (new) legal or public requirements;
- educational offers;
- motivation of employees;
- career planning (Riedel, 2009).

The demand for trainings should be determined systematically and entered in a training plan for the relevant year and then planned accordingly. When carrying out the training, it is also important to document each training verifiably, e.g. as a proof for authorities or within the framework of external audits for certifications. For employees in a company in the food industry, there exists a legal provision to regularly take part in trainings on the subject of 'food hygiene and infection protection law'.

The role of quality management

In the basic standard on the subject of quality management, DIN EN ISO 9001 (see Chapter 5.2), a well-functioning quality management system additionally requires a representative from the top management. This is to be a member of the management being responsible and authorised to carry out the following:

- ensuring that processes needed for the quality management system are established, implemented and maintained;
- reporting to top management on the performance of the quality management system and any need for improvement; and
- ensuring the promotion of awareness regarding customer requirements throughout the organization (DIN EN ISO 9001, 2008).

For small and medium-sized companies and agricultural enterprises an exemption is made. An external person can assume the function of quality management representative. This offers for various network coordinators the possibility to expand their service profile. An entire QM department or various QM departments in different corporate divisions – depending on the size of the company – are subordinate to the quality management representative. This may be one individual employee in small companies. Typical tasks of a quality management representative are the following:

- support and professional guidance for planning and implementation of quality policies and quality objectives;
- communication of quality principles to support quality awareness in companies or the value chain;
- continuous improvement and assessment of the advancement of the quality management system;
- coordination, consulting and training of managers and operative employees in all queries of quality management;
- coordination of preparation, updating and development of QM specifications: work instructions, procedural instructions and other operative documents;
- testing and release of internal/external quality-relevant documents;
- passing on information to the management about quality-related data and coordination of measures for quality improvement;
- assuring of QM supporting documentation (records);
- assistance in assessment of test results and customer surveys;
- coordination of yearly audit program including selection of auditors;
- planning and evaluation of internal audits;

- assuring expert exchange with other institutions in the area of quality, risk and crisis management;
- monitoring of system and product-related corrective and preventive measures;
- supporting the purchasing department regarding qualification and monitoring of suppliers (Figure 9.19).

External quality representatives should be available at the site for a reasonable amount of work time and have the necessary authorities in the company. Frequent tasks include the following: maintenance of the entire documentation, training and briefing of employees (e.g. regarding new QM tools), complete internal audit process, evaluation of complaints including statistics and derivation of measures, providing support to the top management (amongst others for assessment of the management system), test planning, coordination of projects, implementation of quality circles, communication with certification bodies and attending the external audit (Wellems, 2009). Attention has to be paid to the fact that

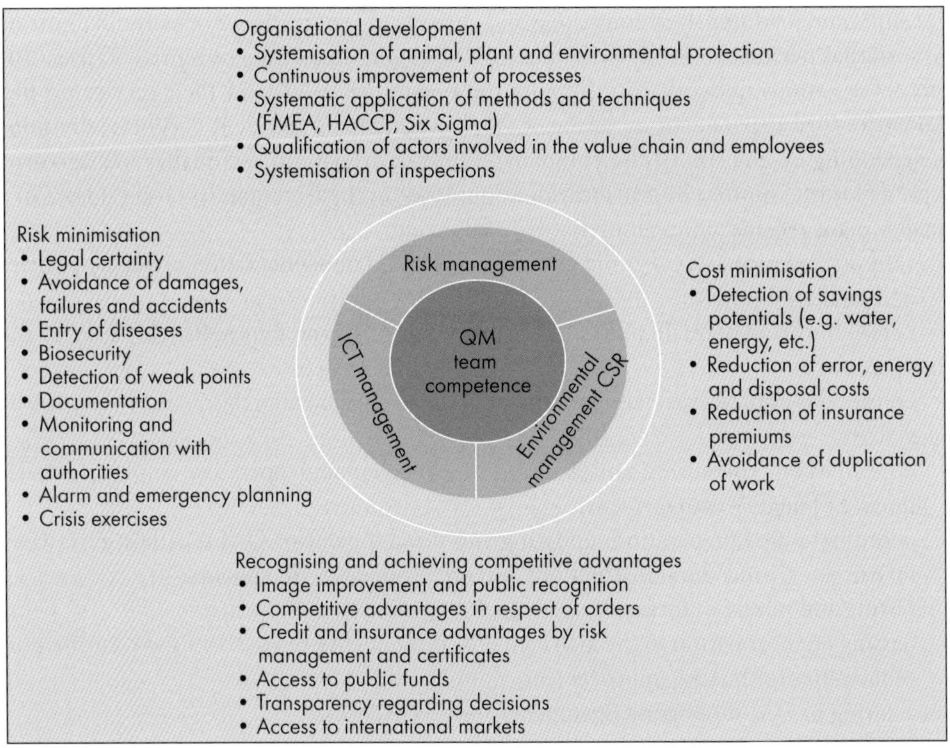

Figure 9.19. Competences of quality management (QM) teams in integrated systems.
HACCP = Hazard Analysis of Critical Control Points; FMEA = Failure Mode and Effect Analysis; CSR = Corporate Social Responsibility; ICT = Information and Communication Technology.

according to DIN EN ISO 9001:2008 this is not the representative of the top management. This role must exist additionally in the company if it has a QM system certified according to DIN EN ISO 9001. Regarding deployment of external employees, the standard says: 'the responsibility of a management representative can include liaising with external parties on matters relating to the quality management system' (DIN EN ISO 9001, 2008; 5.5.2). In integrated quality management systems, a variety of competences has to be provided by the quality representatives to cover the areas of risk management, environmental management, management of information and communication technology, corporate social responsibility, etc. Therefore quality management teams were established.

Confirmation of competence by international certificates

There are several offers on the market in the form of seminars and trainings to qualify an employee to fulfil the task of a quality management representative. This book illustrates as an example systematic training offered by the German Society for Quality (Deutsche Gesellschaft für Qualität e.V.; DGQ) as national representation of the European Organisation for Quality (EOQ) (status: 2013). DGQ offers the relevant qualification to become quality representative and also for other positions in the area of QM independently for each industry. An overview of the qualification scheme of the DGQ in the area of QM is shown in Figure 9.20 (left-hand side). After the participants have completed the two trainings 'Quality Management I – Basic requirements' and 'Quality Management II – Methods, Implementation, Communication' and have taken an oral and written examination regarding the contents of these trainings, they receive the certificate 'DGQ Quality Management Representative/EOQ Quality Management representative '. This certificate documents that the relevant holder is qualified to introduce, maintain and internally audit a quality management system according to DIN EN ISO 9001. And it also proves their know-how on handling the most important tools of quality management.

The next step in systematic further training is 'DGQ quality manager/EOQ Quality Manager'. To get this certificate, the participants have to complete the training courses 'Quality Management II – Didacticals, Analysis, Statistics' and 'Quality Management III – Strategy, Integration, Development' with a final examination. The last step in the qualification scheme of the DGQ is obtaining the auditor qualification 'DGQ Auditor Quality/ EOQ Quality Auditor' (see Chapter 5.3). For this, it is necessary to complete the training course 'Auditor' and successfully pass the examination as well as to demonstrate professional experience and participation in audits. In addition, any other requirements of the relevant system provider must be fulfilled to work subsequently as an external auditor for specific systems such as IFS (International Featured Standard) or QS (Quality and Safety) (see Chapter 5.2). These can usually be found on the internet pages or the corresponding rules and regulations of the system providers.

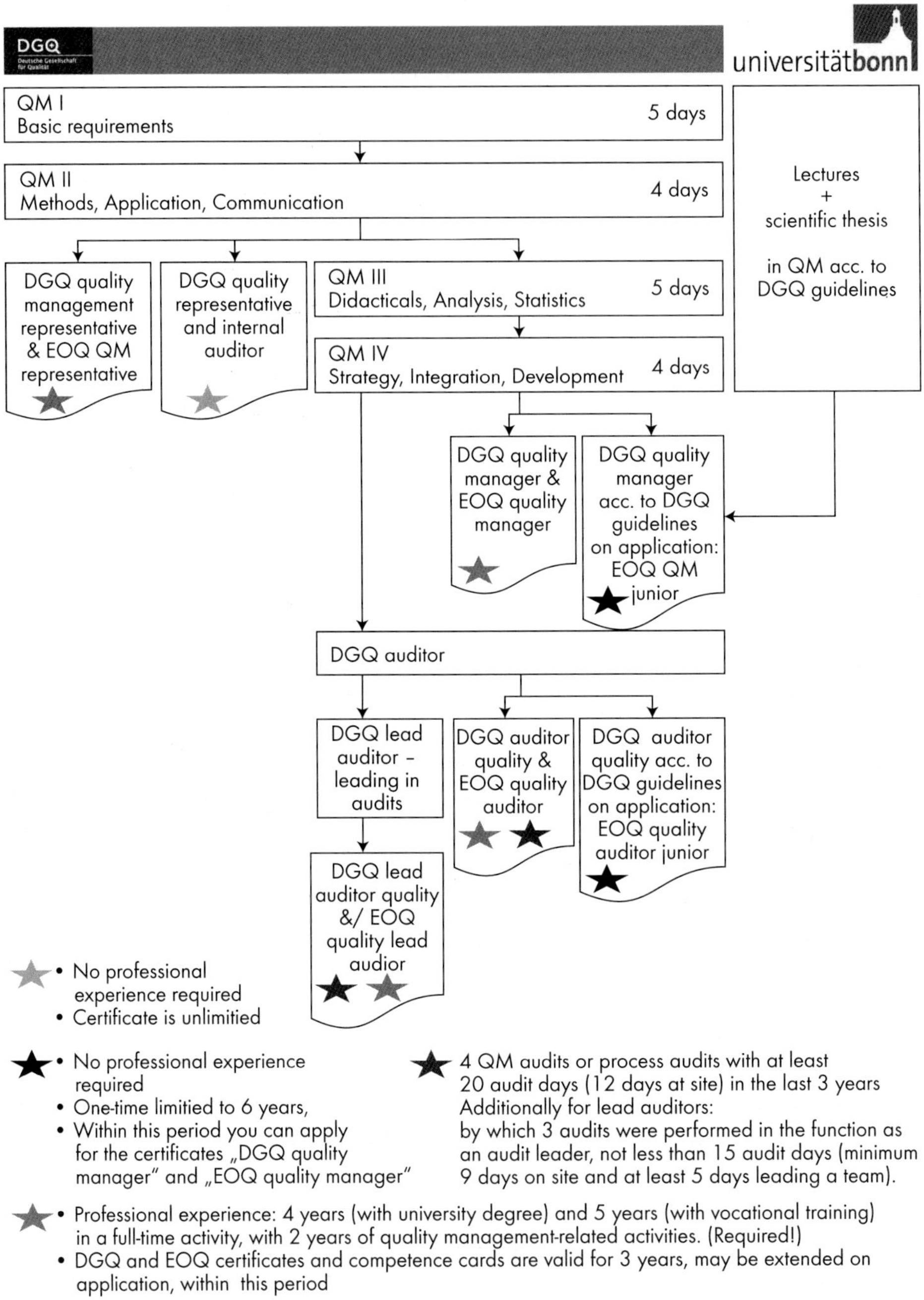

Figure 9.20. DGQ Qualification path in quality management (QM) in combination with EOQ QM Junior program at the University of Bonn.

DGQ = Deutsche Gesellschaft für Qualität; EOQ = European Organisation for Quality; QM = Quality Management.

Students in Germany have the option to acquire additional qualifications in the area of QM recognised by DGQ already during their studies. These additional qualifications are confirmed after successful completion of the studies by a special certificate. Since 1996, DGQ meanwhile offers at 14 universities (status: April 2013, access to current information: www.dgq.de) the qualification model to become 'DGQ Quality Manager acc. to DGQ guidelines/EOQ Quality Manger Junior' (QM Jr.) (Figure 9.20, right-hand side). This model is accompanied by the Beirat Lehre an Hochschulen (BLH, advisory council for university studies). Each year, several hundred students qualify throughout the country. Further training accompanying the studies is mainly chosen by those students who want to gain an advantage for their career in industry or public institutions. The certificate confirms the basic knowledge necessary for further education to become DGQ or EOQ Auditor. This auditor training involves several stages as described above. The additional training to become DGQ Quality Systems Manager Junior enables a lateral career move: the certificate authorises its holder after fulfilling other professional requirements to start immediately with the auditor training to become DGQ Quality/EOQ Quality Auditor. The contents of the additional training correspond with the requirements of the harmonised education scheme of the EOQ and the standard EOQ CoS 9001 Edition 2012[11].

A quality manager is responsible for the entire process-oriented quality management system and for its establishment and strategic further development. In this regard, he considers key performance indicators and acts systematically and in a structured way to make the processes of the organisation transparent. The quality manager supports effective corporate communication and has a bridging function across several areas. These management tasks require a high degree of expertise, communication skills and personal integrity. On the way to become DGQ Quality Manager acc. to DGQ guidelines/EOQ Quality Manger Junior, it is necessary to acquire the required competences and apply them accordingly, such as

- understanding of the fundamentals of quality management;
- principles of introduction, maintenance and advancement of a quality management system;
- organisation of internal audits;
- application of quality management tools;
- taking decisions on the basis of statistical data;
- designing of processes in quality management towards total quality management and continuous improvement.

This expert knowledge is conveyed within the framework of lectures, exercises, practical training and workshops. In addition to this, practical application is done by means of moderation, presentation, teamwork and demonstration of implementation of systems in the agri-food industry. The advantages of the certificate are as follows:

[11] www.eoq.org/eoq_competence_specifications_and_certification_schemes/request_eoq_cs_9000.html.

- certification of extensive knowledge in the field of quality management;
- additional qualification for starting the professional career;
- time and cost saving in comparison with the non-university education to become 'DGQ Auditor Quality/EOQ Quality Auditor'.

To acquire an additional qualification, students have to attend lectures and take examinations in three different blocks at any university having an accredited DGQ/EOQ training model:
- Block I Lectures and exercises (at least 100 hours);
- Block II Seminars and teamwork (at least 40 hours);
- Block III Practical work (at least 200 hours).

All three areas are covered by special modules/subjects with the relevant written or oral examinations during the studies determined in coordination with the BLH.

The results achieved during empirical studies in the agri-food industry show clearly that quality management has become more dynamic in this respect as well during recent years. QM departments and quality management representatives feeling responsible for the entire value chain, today, need to be equally far sighted and secure when it comes to taking prompt decisions. It is necessary, on the one hand, to counter the continuously changing market requirements by designing long-term strategies and, on the other hand, by taking decisions in crisis situations very quickly (Petersen, 2012b).

Team spirit and the willingness to enter into cooperations are nowadays amongst the outstanding qualities characterising junior employees for the top management of tomorrow. Those who do not learn already during their university education to think and work in interdisciplinary terms will have difficulties during their professional career to acquire these competences. Therefore, an alliance meanwhile operating internationally consisting of universities and industrial organisations is aiming at supporting education, mobility and career development of junior staff. In Secion 9.3, the alliance ALFONS was already introduced. In this connection it is important to find dynamic solutions for the exchange of personnel between public research institutions and companies, especially small and medium-sized enterprises, including traditional associations and industrial stakeholders. The exchange of knowledge and mutual understanding of differing framework conditions and qualification requirements in academia and industry are the basis for advancement of personnel.

New fields of activity and more complex tasks in the area of agri-food industry require a qualified training of experts. In addition to sound expert knowledge, special skills, amongst others in quality management, communication and risk and crisis management are necessary. In international university networks, contents are already conveyed in the study program 'Food Safety Management'. In the project quarisma (QUAlity and

RISk Management in meat chains), a quality management training guided by European standards and accompanying the studies for future ecotrophologists, food technologists and agro-scientists is developed. Cross-departmental interdisciplinary interaction of many network partners is necessary for the DGQ Quality Manager acc. to DGQ guidelines/ EOQ Quality Manger Junior program designed by EOQ and DGQ. Educational partners in the university and private sector and also companies and organisations of the agri-food industry have discovered competent experts and managers as their key resource.

The education program is further developed within the framework of cooperation of universities and representatives of EOQ and DGQ to fulfil the extended range of tasks of quality and risk management in future as well. It is planned to award the certificate 'Quality Systems and Risk Manager Junior (QRM Jr.)' in future.

Global markets necessitate strategic orientation towards European and international standards when extending the career perspectives in quality management of industry and academia. Therefore, the intended QM training is further developed jointly with the European quality management networks of the EOQ. The EOQ functioning as the umbrella organisation of selected national quality societies suggests harmonised education and certification concepts for quality and management personnel.

Training courses and seminars initiated by the quarisma project are targeted towards employees of companies and organisations. The QM personnel working there can realise its additional qualification and refreshing of knowledge in these courses which are open to academics from industry. The industrial partners make sure that the focus of such training components is oriented contextually towards the requirements of quality management of companies and organisations of the value chain.

Chapter 10. Epilogue

M. Hamer[1], B. Petersen[2], M. Nüssel[3]
[1]GIQSe.V., c/oUniversityofBonn, Katzenburgweg 7-9, 53115Bonn,Germany
[2]University of Bonn, Katzenburgweg 7-9, 53115Bonn, Germany
[3]DRV, DeutscherRaiffeisenverband e.V., Pariser Platz 3, 10117 Berlin, Germany
m.hamer@giqs.org

It is estimated that the growing global population, increasing prosperity of the emerging countries and modified consumption habits will double the demand for food up to the year 2050. It will be challenging to meet future global demands for food while avoiding, or at least mitigating, negative impacts on ecosystems – and at the same time reducing poverty, supporting livelihoods, and ensuring food safety and animal welfare (UNEP 2012). At the same time, we are confronted with changing consumer demands, the challenges of climate change and advancing globalisation as well as shortages of resources (Schwerin *et al.*, 2010). Other assessments indicate that nowadays nearly half of our food is already lost before it reaches private households. The reasons for this lie, amongst others, in rotting and destruction by parasites already in the field, inefficient management of supply chains, weak points in the management of cold chains and inadequate local infrastructures (J. Matern, personal communication). The global challenge to produce more food and to ensure, at the same time, that this food is of high quality, in consideration of the principle of sustainability, therefore requires research approaches and management strategies guaranteeing an overall view and interconnection of all links and players of the feed and food chains in an international context. Quality sciences as a cross-sectional discipline can contribute in this regard enormously to responsible and sustainable food production.

Quality and risk management are – as shown in this book – very complex issues examined by the authors from a scientific and also from a practical point of view and are explained by way of concrete examples. Hence, they classify the most important concepts of quality management, describe the development stages from quality control right up to the dynamic quality management network in global value chains and show which models are used nowadays to illustrate interdependencies and the tasks of those responsible for quality and risk management. The framework provided for establishing QM systems in individual operations or within value chains according to the 'farm to fork' or 'fork to farm' principle

DOI 10.3920/978-90-8686-789-9_10, © Wageningen Academic Publishers 2014

is, on the one hand, EU food legislation, and, on the other hand, international rules and standards. The latter are, to a large extent, independent of the relevant branch of business and require sector-specific interpretation assistance for their implementation.

Four current fields of action will advance the continuous improvement process within the QM systems of international value chains:

- increasing the visibility of successful quality and risk management systems;
- improving the business-to-business communication;
- intensifying the dialog between industry and academia;
- strengthening of networks and public-private partnership structures.

The experience reports mentioned in this book show clearly that quality management has become more dynamic also in the agri-food industry. QM departments or quality management representatives feeling responsible for the entire value chain, nowadays have to be far sighted and able to react any time as well as confident and prudent when taking rapid decisions in situations of crisis. For this, it is necessary, on the one hand, to establish long-term structures and thus countering the continuously changing market demands, and be well prepared, on the other hand, for situations requiring decisions in crises.

The demand for tailored further training programs for future quality managers in industry and know-how transfer from research have therefore continuously increased in recent years. Thus, for example, the initiative ALFONS (Alliance for training and further training FoodNet Systems) explains clearly which development routes can be followed to offer corresponding qualification opportunities during and after studies to junior employees and managers in the agri-food industry. Meanwhile, the scope of quality managers also includes advancement of quality management in global company structures and integration of quality, risk, environmental and energy management. Dynamic markets require dynamic QM systems, perfectly coordinated processes, integrated IT systems across company borders and clear and rapid de-escalation strategies.

Another response to globalisation is supplier development. Here, additional knowledge of industry-specific QM methods and procedures is necessary. Creation of new jobs within the entire value chain reflects this development and shows the demand for well-educated junior employees.

Even customers and consumers contribute a certain dynamics to quality management. The quality of future products will be increasingly influenced by communication with the consumer. It is necessary to capture the requirements of customers and consumers by targeted enquiries and subsequently to communicate the fulfilment of customer requests.

Value chains in the agri-food industry are characterised by geographically and organisationally separated process steps resulting in information but also responsibility asymmetries. For the future it can be predicted that there will be an increasing willingness to not only exchange important product and process information with the companies of the value chain rapidly and suited to the demand, but also with public institutions. Already today, a number of exemplary projects illustrate how B-to-B and PPP communication via the internet can be realised organisationally and technically. In global quality networks, innovation in the public and private sector plays an increasingly important role. The public sector at national, regional and local level is under pressure to create new ways of offering better public goods and services with fewer resources. The rapid pace of innovation in many fields of quality, risk and crisis management creates tremendous opportunities for improving the cost effectiveness and impact of the public sector, including new ways of cooperation with the private sector. Novel public-private partnerships play an important role in knowledge-based development and social entrepreneurship.

The interface between information and communication systems of network coordinators on the agricultural side and DP systems of companies such as slaughterhouses, dairies, mills, breweries, etc. at the processing stage will also be in the centre of joint projects between industry and academia in the coming years. Advancement concepts for information and service agencies at the producer stage are in trend in this respect to increase transparency and competitiveness of the entire value chain.

Quality and safety of agricultural products and food has utmost priority for consumers worldwide. With this highly sensitive purchaser market, organisation and certification of quality communication during production and across the stages remain one of the most important success factors for competing with international competitors and developing new markets. This complex subject is associated with a high need for coordination not only between the partners of the value chain but also with academia. The best way to exploit the potential of different innovation activities with several partners of a value chain is through structured planning and control of innovation activities from the initial idea to successful market entry. To facilitate this dialog, in the past years, R&D platforms like GIQS have been developed and scientists have taken over neutral moderation of project teams with partners from agriculture to food retail. According to the network partners involved, tackling research and development issues jointly will time and again result in a number of advantages and synergy effects.

If all these initiatives and measures succeed to establish quality and sustainability as key factors in added chains and to advance organisations by training junior employees and managers necessary for this, quality will remain a central differentiation characteristic in global markets in future as well. However, we must not forget that in addition to networking and communication skills, not least motivation and pioneering spirit, courage for innovation and intercultural competence are part of a holistic quality understanding in the agri-food industry.

References

Adler, P.S. and Borys, B., 1996. Two types of bureaucracy: enabling and coercive. Administrative Science Quarterly 41, 61-89.

Ahlert, D. and Schulze-Bentrop, C., 2010. Pricing of solutions. Projektbericht Transolve. Universität Münster, Münster, Germany.

AID, Infodienst Verbraucherschutz, Ernährung, Landwirtschaft e.V., 2003. Zentrale Datenbank – jetzt auch für Schweine, Issue No. 1493, Bonn, Germany.

Akerlof, G.A., 1970. The market for 'lemons': quality uncertainty and the market mechanisms. Quarterly Journal of Economics, 84th year, 488-500.

Anderssen, T., 1995. Zertifizierung nach ISO 9000 in der Landwirtschaft. In: FCL-Schriftenreihe Band 4 "DIN EN ISO 9000 ff.: Neue Impulse für das Qualitätsmanagement in der europäischen Agrarwirtschaft". Landwirtschaftsverlag, Münster, Germany, 13 p.

Andrei, P., 2005. Praxishandbuch Hygiene und HACCP – Leitfaden mit Arbeitshilfen für die Lebensmittelwirtschaft. Behr's Verlag, Hamburg, Germany.

Apelt, M., 1999. Vertrauen in der zwischenbetrieblichen Kooperation. Deutscher Universitätsverlag, Wiesbaden, Germany.

Arens, L., Deimel, M. and Theuvsen, L., 2010. Transparenz in der Fleischerzeugung – Wahrnehmung durch den Verbraucher am Point of Sale. Yearbook of Socioeconomics in Agriculture 4, 189-216.

Arens, L., Deimel, M. and Theuvsen, L., 2011. Qualitätskommunikation am Point of Sale – Ergebnisse einer adaptiven Conjoint-Analyse. In: Petersen, B. and Raab, V. (eds.) Qualitätskommunikation. Shaker Verlag, Aachen, Germany, pp. 29-51.

Arens, L., Plumeyer, C.-H. and Theuvsen, L., 2012. Determinants of the use of information: an empirical study of German pig farmers. International Food and Agribusiness Management Review 15(1), 51-72.

Arnold, U., 1997. Beschaffungsmanagement. Schäffer-Poeschel Verlag, Stuttgart, Germany.

Aslesen, H.W., Braadland, T.E., Jensen, L.H., Isaksen, A. and Ørstavik, F., 1999. Innovation, knowledge bases and clustering in selected industries in the Oslo region. Stiftelsen STEP Report 4, Oslo, Norway.

Azevedo, S.G., Carvalho, H., Cruz Machado, V., 2011. The influence of green practices on supply chain performance: a case study approach. Transportation research: an international journal 47, 850-871.

DOI 10.3920/978-90-8686-789-9_10, © Wageningen Academic Publishers 2014

Bahlmann, J. and Spiller, A., 2008. Innovative Lieferantenbewertungsinstrumente zur Steigerung der Wettbewerbsfähigkeit der deutschen Schweinefleischproduktion. In: Spiller, A. and Schulze, B. (eds.) Zukunftsperspektiven der Fleischwirtschaft: Verbraucher, Märkte, Geschäftsbeziehungen. Universitätsverlag Göttingen, Germany, pp. 97-130.

Bahlmann, J. and Spiller, A., 2009. Inter-organisational information systems in meat chains: the linkage between supply chain organisation and system requirements. Journal on Chain and Network Science 9(1), 59-69.

Baptista, F.M., Dahl, J. and Nielsen, L.R., 2010. Factors influencing Salmonella carcass prevalence in Danish pig abattoirs. Preventive Veterinary Medicine, 95(3), 231-238.

Baroni, B., Baroni, M.R. and Torri, L., 2009. Modified atmosphere. Artek Publishing, Food Packages Free Press, Varese, Italy.

Bassewitz, H. v. Graf, 2002. Verbund – Fessel oder Chance? In: DLG (ed.) Landwirtschaft in der Ernährungswirtschaft. DLG-Verlag, Vol. 96, Frankfurt am Main, Germany, pp. 41-48.

Batterink, M., Wubben, E.F.M. and Omta, S.W.F., 2006. Factors related to innovative output in the Dutch agrifood industry. Journal on Chain and Network Science 6(1), 31-44.

Bell, R.G., 2005. Meat packaging: protection, preservation and presentation. Meat Science and Applications. 1st ed., Hui, Y.H. (ed.) Marcel Dekker Inc., New York, NY, USA, 463 pp.

Berns, G., 1996. Einbindung von Check-Listen und mobilem Analyselabor in Beratungskonzepte zur Erweiterung von Gesundheitsvorsorge- und Qualitätsmanagement in der Schweineerzeugung. Dissertation Universität Bonn, Bonn, Germany.

Beulens, A.J.M., 2003. Transparency requirements in supply chains and networks: yet another challenge for the business and ICT community. In: Ehrenberg, D. and Kaftan, H.J. (eds.) Herausforderungen der Wirtschaftsinformatik in der Informationsgesellschaft. Wirtschaftsverlag, Leipzig, Germany, pp. 203-223.

Beulens, A.J.M., Broens, D.F., Folstar, P. and Hofstede, G.J., 2005. Food safety and transparency in food chains and networks relationships and challenges. Food Control 16(6), 481-486.

Bijman, J., Omta, S.W.F., Trienekens, J.H., Wijnands, J.H.M. and Wubben, E.F.M. (eds.), 2006. International agri-food chains and networks: management and organisation. Wageningen Academic Publishers, Wageningen, the Netherlands.

BioSt-NachV, 2009. Verordnung über Anforderungen an eine nachhaltige Herstellung von flüssiger Biomasse zur Stromerzeugung Biomassestrom-Nachhaltigkeitsverordnung (BioSt-NachV) vom 23.07.2009, Bundesministeriums der Justiz in Zusammenarbeit mit juris GmbH, Saarbrücken. Germany. Available at: www.gesetze-im-internet.de/bundesrecht/biost-nachv/gesamt.pdf.

Blanke, W., 2002. Hand in hand – rating und risikomanagement können qualitätsmanagement und controlling zusammenwachsen lassen. Qualität und Zuverlässigkeit 47(5), 490-491.

Blokhuis, H.J., Keeling, L.J., Gavinelli, A. and Serratotosa, J., 2008. Animal welfare's impact on the food chain. Trends in Food Science & Technology, 19, 79-87.

Boon, W.P.C, 2008. Demanding dynamics. Demand articulation of intermediary organisations in emerging pharmaceutical innovations. Dissertation. Utrecht University, Utrecht, the Netherlands.

Brans, J.P., Vincke, P. and Mareschal, B., 1986. How to select and how to rank projects: The PROMETHEE method. Journal of Operational Research 24(2), 228-238.

Breitschuh, G. and Gerhard, M., 2012. Wir produzieren nachhaltig. In: dlz agrar magazin, Mai 2012. Deutscher Landwirtschaftsverlag, Munich, Germany, pp. 26-32.

Brenner, P.-F., Akkasoglu, G. and Weckmann, A., 2010. Rasch und sicher auf neue Wege – Reifegradmethode für neue Fertigungsverfahren. Qualität und Zuverlässigkeit 55(6), 52-55.

Breuer, O., Saatkamp, H., Schütz, V., Brinkmann, D. and Petersen, B., 2008. Cross border classical swine fever control: improving Dutch and German crisis management systems by an integrated public-private approach. Journal für Verbraucherschutz und Lebensmittelsicherheit 3(4), 455-465.

Brinkmann, D. and Petersen, B., 2010. Qualitätsverantwortung in der Wertschöpfungskette. In: Petersen, B., Spiller, A. and Theuvsen, L. (eds.) Vom Viehvermarkter zum Dienstleistungsprofi. GIQS e.V., Bonn, Germany, pp. 15-21.

Brinkmann, D. and Petersen, B., 2011. Nachhaltigkeit in der Wertschöpfungskette. Stufenübergreifende Management-Maßnahmen sind in der Fleischbranche noch Mangelware. Fleischwirtschaft 3, 28-30.

Brinkmann, D., Lang, J., Petersen, B., Wognum, N. and Trienekens, J., 2011. Towards a chain coordination model for quality management strategies to strengthen the competitiveness of European pork producers. Journal on Chain and Network Science 11, 137-153.

Brody, A.L., Strupinsky, E.R. and Kline, L.R., 2001. Active packaging for food applications. Technomic Publishing Company, Lancaster, PA, USA.

Bruckner, S., Albrecht, A., Petersen, B. and Kreyenschmidt, J., 2013. A predictive shelf life model as a tool for the improvement of quality management in pork and poultry chains. Food Control 29, 451-460.

Brühwiler, B., 2009. Die Norm ISO 31000 – MQ-Serie Risikomanagement (Teil 1 Grundlagen). MQ – Management und Qualität 1-2, 24-26.

Bruns, M., 2011. Inter-organisational innovation processes in the agrifood industry: An approach to improving management support services applied to the meat industry. Dissertation. Universität Bonn, Bonn, Germany.

Bruns, M., Petersen, B. and Maijers, W., 2008. Intermediary act as research-industry liaison office. Support in structuring the innovation process especially for small and medium sized enterprises of the pork chain. Proceedings of the 8th International Conference on Management in AgriFood Chains and Networks. 28-30 May 2008. Ede, the Netherlands.

Buhl, C.M., 2009. Erhöhung der Innovationskraft durch Kooperationen in Netzwerken und Clustern. In: Bundesministerium für Wirtschaft und Technologie (ed.) Innovative Netzwerkservices. Netzwerk- und Clusterentwicklung durch maßgeschneiderte Dienstleistungen. Berlin, Germany, pp. 13-20.

Bundesministerium für Umwelt, Naturschutz und Reaktorsicherheit (BMU), 2010. Nachhaltigkeit braucht Führung: bewusst – kompetent – praxisnah. Available at: www.bmu.de/files/pdfs/allgemein/application/pdf/broschuere_csr_nachhaltigkeit_fuehrung.pdf.

Buranasompob, A., 2005. Kinetics of the inactivation of microorganisms by water insoluble polymers with antimicrobial activity. Dissertation, Technische Universität Berlin, Berlin, Germany.

Chesbrough, H., 2003. Open Innovation. The new imperative for creating and profiting from technology. Boston, MA, USA.

Christopher, M. and Peck, H., 2004. Building the resilient supply chain. International Journal of Logistics Management 15, 1-13.

Church, N., 1994. Review - Developments in modified-atmosphere packaging and related technologies. Trends in Food Science and Technology 5, 345-352.

Codex Alimentarius Commission 1999. Report of the thirtieth session of the Codex Committee on pesticide residues, The Hague, 20 - 25 April 1998. FAO/WHO, Rome, Italy. Available at: http://www.google.de/url?sa=t&rct=j&q=&esrc=s&source=web&cd=4&ved=0CFAQFjAD&url=http%3A%2F%2Fwww.codexalimentarius.org%2Finput%2Fdownload%2Freport%2F201%2FAl99_24e.pdf&ei=woCoUqyYPMWByQP9lYDIDg&usg=AFQjCNGcN9vbMb4kKr7TEnzYnk--VdIwVA&sig2=0DoCsnirmqypsm5NnL_AMg&bvm=bv.57799294,d.bGQ.

Codex Alimentarius Commission, 1997. Recommended international code of practice – general principles of food hygiene. FAO, Rome, Italy. Available at: http://www.fao.org/docrep/w8088e/w8088e04.htm#module%201%20%20%20the%20codex%20alimentarius%20commission.

Codex Alimentarius Commission, 2003. Recommended international code of practice - general principles of food hygiene including annex on hazard analysis and critical control point (HACCP) system and guidelines for its application (CAC/RCP 1-1969, Rev. 4-2003 – Annex). Ministry of Health, Labour and Welfare, Tokyo, Japan. Available at: http://www.mhlw.go.jp/topics/yunyu/siryo/dl/siryo02l.pdf.

Collins, J.D. and Wall, P.G., 2004. Food safety and animal production systems: controlling zoonoses at farm level. Revue scientifique et technique (International Office of Epizootics) 23, 685-700.

Coma, V., 2008. Bioactive packaging technologies for extended shelf life of meat-based products. Meat Science 78, 90-103.

Cooksey, K., 2001. Antimicrobial food packaging materials. Additives for Polymers, 2001(8) 6-10.

Czarnitzki, D., Licht, G., Rammer, C. and Spielkamp, A., 2001. Rolle und Bedeutung von Intermediären im Wissens- und Technologietransfer. ifo Schnelldienst 54(4), 40-49.

De Giovanni, P. and Esposito Vinzi, V., 2012. Covariance versus component-based estimations of performance in green supply chain management. International Journal of Production Economics 135, 907-916.

De Marchi, V., 2012. Environmental innovation and R&D cooperation: Empirical evidence from Spanish manufacturing firms. Research Policy 41(3), 614-623.

De Swarte, C. and Donker, R.A., 2005. Towards an FSO/ALOP based food safety policy. Food Control 16, 825-830.

Deimel, M., Frentrup, M. and Theuvsen, L., 2008a. Transparency in food supply chains: empirical results from German pork and dairy production. Journal on Chain and Network Science 8(1), 21-32.

Deimel, M., Plumeyer, C.-H. and Theuvsen, L., 2009. Stufenübergreifender Informationsaustausch in der Fleischwirtschaft: Recht und Zertifizierung als Einflussgrößen. Berichte über Landwirtschaft 87, 118-152.

Deimel, M., Theuvsen, L. and Ebbeskotte, C., 2008b. Von der Wertschöpfungskette zum Netzwerk: Methodische Ansätze zur Analyse des Verbundsystems der Veredelungswirtschaft Nordwestdeutschlands, Discussion paper No. 0810. Departemnt of Agricultural Economics and Rural Development, Universität Göttingen, Göttingen, Germany.

Den Hertog, P. and Roelandt, T.J.A., 1999. Cluster analysis and cluster-based policy making: the state of the art. In: OECD – Organisation for Economic, Co-operation and Development: Boosting innovation. The cluster approach. Paris, France, pp. 413-427.

Derlien, H.-U., 1992. Bürokratie. In: Frese, E. (ed.) Handwörterbuch der Organisation. 3rd Edition, Schäffer-Poeschel Verlag, Stuttgart, Germany, pp. 391-400.

Deutsche Gesellschaft für Qualität e.V.(DGQ), 1998. Qualitätsmanagement für kleine und mittlere Unternehmen der Investitionsgüterindustrie. Forschungsbericht/Forschungsgemeinschaft Qualität e.V. (FQS) Beuth Verlag, Berlin, Germany.

Deutsche Gesellschaft für Qualität e.V.(dgQ), 2000. Qualitäts- und Effektivitätssteigerung bei F&E-Projekten durch methodenunterstütztes Projektmanagement. FQS-DGQ-Band 86-01. Frankfurt am Main, Germany.

DGQ-Schrift 11-04: Begriffe zum Qualitätsmanagement. Beuth Verlag, Berlin, Germany.

DIN DIN EN 45020, 2007-03. Normung und damit zusammenhängende Tätigkeiten – Allgemeine Begriffe (ISO/IEC Guide 2:2004); Trilingual version EN 45020:2006.

DIN EN ISO 19011:2011-12 Leitfaden zur Auditierung von Managementsystemen, Beuth Verlag, Berlin, Germany.

DIN EN ISO 22000, 2005 Managementsysteme für Lebensmittelsicherheit - Anforderungen an Organisationen in der Lebensmittelkette, Beuth Verlag, Berlin, Germany.

DIN EN ISO 50001, 2011-12 Energiemanagementsysteme - Anforderungen mit Anleitung zur Anwendung, Beuth Verlag, Berlin, Germany.

DIN EN ISO 9000, 2005 Qualitätsmanagementsysteme – Grundlagen und Begriffe, Beuth Verlag, Berlin, Germany.

DIN EN ISO 9001, 2008 Qualitätsmanagementsysteme – Anforderungen (ISO 9001:2008); Trilingual version EN ISO 9001:2008. Deutsches Institut für Normung e.V., 2008, Berlin, Germany.

DIN EN ISO 9004, 2009: Managing for the sustained success of an organisation – a quality management approach (ISO 9004:2009). Trilingual version EN ISO 9004:2009. Beuth Verlag, Berlin, Germany.

DIN ISO 14001, 2012. Überarbeitung der Umweltmanagement-Norm ISO 14001. Available at: http://www.din.de/cmd?level=tpl-artikel&menuid=47387&cmsareaid=47387&cmsrubid=47393&menurubricid=47393&cmstextid=158445&2&languageid=de.

DIN ISO 26000, 2011. Leitfaden zur gesellschaftlichen Verantwortung. Beuth Verlag, Berlin, Germany.

DIN ISO 31000, 2009. Risk management – Principles and guidelines (ISO 31000:2009), 2011. Beuth Verlag, Berlin, Germany.

DIN ISO 26000, 2010. Guidance on social responsibility, Beuth Verlag, Berlin, Germany.

DiPiazza, S.A. and Eccles, R.G., 2002. Building public trust. Wiley, New York, NY, USA.

Directorate General for Health & Consumers, European Commission (DG SANCO), 2006. Labelling: competitiveness, consumer information and better regulation for the EU. Consultative Document. DG SANCO, Brussels, Belgium.

Doluschitz, R., 2001. Kooperationen in der Landwirtschaft. Berichte über Landwirtschaft 79, 375-398.

Drescher, K., 1993. Vertraglich vertikale Koordination in der deutschen Landwirtschaft. Dissertation, Universität Kiel, Kiel, Germany.

DS49001, 2011. Managementsystem für gesellschaftliche Verantwortung - Anforderungsbeschreibung, Beuth Verlag, Berlin, Germany.

Düsseldorf, S., 2013. Concept of key performance indicators controlling consumer oriented quality and herd health management in a Bavarian pork chain. Dissertation, Universität Bonn, Bonn, Germany.

Düsseldorf, S., Kagerer, C., Petersen, B. and Brinkmann, D., 2011. Qualitätskommunikation in einem regionalen Netzwerk. In: Petersen, B. and Raab, V. (eds.) Berichte zum Qualitätsmanagement – Qualitätskommunikation. GQW Volume 12, Bonn, Germany, pp. 75-96.

Eden, M., Raab, V., Kreyenschmidt, J., Haflidason, T., Ólafsdottir, G. and Bogason, S.G., 2011. Continuous temperature monitoring along the chilled food supply chain. In: Hoorfar, J., Jordan, K., Butler, F. and Prugger, R. (eds.) Food chain integrity: a holistic approach to food traceability, safety, quality and authenticity. Woodhead Publishing, Sewston, Cambridge, UK.

Ellebrecht, A., 2008. Nutzenbetrachtung internetbasierter Informationssysteme im einzel- und überbetrieblichen Gesundheitsmanagement. Dissertation, Universität Bonn, Bonn, Germany.

Ellebrecht, S., 2012. Anwendungskonzept kombinierter präventiver Qualitätsmanagement-Methoden im überbetrieblichen Gesundheitsmanagement Schweinefleisch erzeugender Ketten. Dissertation, Universität Bonn, Bonn, Germany.

Ellebrecht, S., Brinkmann, D. and Petersen, B., 2009: Quality management systems in European pork chains. In: Trienekens, J., Petersen, B., Wognum, N. and Brinkmann, D. (eds.) European pork chains. Wageningen Academic Publishers, Wageningen, the Netherlands, p. 75-92.

Ercolini, D., Ferrocino, I., Storia, A.L., Mauriello, G., Gigli, S., Masi, P. and Villani, F., 2010. Development of spoilage microbacter in beef stored in nisin activated packaging. Food Microbiology 27, 137-143.

Erich, M., 2012. Food 2030: Samenwerking vanuit een nieuwe mindset. ING, the Netherlands, 59 p.

EU Commission, 2009. Communication from the Commission to the European Parliament, the Council, the European Economic and Social Committee and the Committee of the Regions. COM/2009/0234 final, Brussels, Belgium.

EU, 2002. Regulation (EU) No 178/2002 of the European Parliament and of the Council of 28 January. Available at: http://www.fsai.ie/uploadedFiles/Legislation/Food_Legisation_Links/General_Principles_of_Food_Law/Consol_Reg178_2002.pdf.

EU, 2004a. Regulation (EC) No 852/2004 of the European Parliament and of the Council of 29 April 2004 on the hygiene of foodstuffs. Official Journal of the European Union L 139, 22-82.

EU, 2004b. Regulation (EC) No 853/2004 of the European Parliament and of the Council of 29 April 2004 laying down specific hygiene rules for food of animal origin. Official Journal of the European Union L 139, 55-206.

EU, 2004c. Regulation (EC) No 854/2004 of the European Parliament and of the Council of 29 April 2004 laying down specific rules for the organisation of official controls on products of animal origin intended for human consumption. Official Journal of the European Union L 226, 83-127.

EU, 2005. Regulation No 2073/2005 of the European Parliament and of the Council of 15 November 2005 on microbiological criteria for foodstuffs. Official Journal of the European Union L 338, 1-26.

EU, 2006. Regulation (EC) No 1924/2006 of the European Parliament and of the Council of 20 December 2006 on nutrition and health claims made on foods. Official Journal of the European Union L 404, 9-25.

EU, 2007. Regulation (EC) No 834/2007 of 28 June 2007 on organic production and labelling of organic products and repealing Regulation (EEC) No 2092/91. Official Journal of the European Union L 189, 1-23.

EU, 2008. Regulation (EC) No 765/2008 of the European Parliament and of the Council of 9 July 2008 setting out the requirements for accreditation and market surveillance relating to the marketing of products and repealing Regulation (EEC) No 339/93. Official Journal of the European Union L 218, 30-48.

EU, 2011. Regulation (EU) No 1169/2011 of the European Parliament and of the Council of 25 October 2011 on the provision of food information to consumers. Official Journal of the European Union L 304, 18-63.

EU, 2012a. Regulation (EC) No 1151/2012 of the European Parliament and of the Council of 21 November 2012 on quality schemes for agricultural products and foodstuffs. Official Journal of the European Union L 343, 1-29.

EU, 2012b. Regulation (EC) No 1893/2006 of the European Parliament and of the Council of 20 December 2006 establishing the statistical classification of economic activities NACE Revision 2 and amending Council Regulation (EEC) No 3037/90 as well as certain EC Regulations on specific statistical domains. Official Journal of the European Union L 393, 1-39.

European Commission (EC), 2000. White paper on food safety. European Commission, Brussels, Belgium. Available at: http://ec.europa.eu/food/food/intro/white_paper_en.htm.

European Communities, 2008. Insights on innovation management in Europe. Tangible results from IMP3rove. European Communities and Europe Innova. Luxembourg, Luxembourg.

Eurostat, 2010. Europa in Zahlen. Eurostat Jahrbuch 2010. Amt für Veröffentlichungen der Europäischen Union, 2010 Luxemburg, Luxembourg, doi: 10.2785/40621.

Floros, J. D. and Matsos, K.I., 2005. Introduction to modified atmosphere packaging. In: Han, J. (ed.) Innovations in food packaging. Elsevier Academic Press, Amsterdam, the Netherlands, pp. 151-171.

Folaron, J., 2003. The evolution of six sigma. Six Sigma Forum Magazine 2(4), 38-44.

Food and Agriculture Organisation of the United Nations (FAO), 2008. Risk-based food inspection manual Rome, FAO Food and Nutrition Paper pp. 89-92. Available at: http://www.fao.org/docrep/010/i0096e/i0096e00.htm.

Food and Agriculture Organisation, OIE World Organisation for Animal Health, International Bank for Reconstruction and Development, 2010. Good practices for biosecurity in the pig sector: Issues and options in developing and transition countries (Repr. 2010 (July)). FAO animal production and health paper: Vol. 169. Rome, Italy. Retrieved from http://www.fao.org/docrep/012/i1435e/i1435e00.pdf.

Fortuin, F., Omta, S.W.F. and Batterink, M., 2007. Key success factors of innovation in multinational agrifood prospector companies. International Food and Agribusiness Management Review, 10(4), 1-24.

Fraqueza, M.J. and Barreto A.S., 2009. The effect on turkey meat shelf life of modified-atmosphere packaging with an argon mixture. Poultry Science 88, 1901-1998.

Freeman, C., 2002. Continental, national and sub-national innovation systems – complementarity and economic growth. Research Policy 31(2), 191-211.

Frentrup, M. and Festag, R., 2012. Risikomanagement in der Fleischwirtschaft: Herausforderungen und Bedeutung einer Verbundstrategie für ein mittelständisches Viehvermarktungsunternehmen. In: Theuvsen, L., Frentrup, M. and Emmann, C.H. (eds.) Risikomanagement in Agrarhandel und Lebensmittelindustrie. Agrimedia, Clenze, Germany, pp. 185-199.

Frentrup, M. and Theuvsen, L., 2006: Transparency in supply chains: is trust a limiting factor? In: Fritz, M., Rickert, U. and Schiefer, G. (eds.) Trust and risk in business networks. ILB-Press, Bonn, Germany, pp. 65-74.

Frentrup, M., Heyder, M. and Theuvsen, L., 2010. Risikomanagement in der Landwirtschaft. Landwirtschaftliche Rentenbank, Frankfurt a. Main, Germany.

Gampl, B., 2006. Rückverfolgbarkeit von Lebensmitteln – Eine empirische Analyse kettenübergreifende Informationssysteme. Cuvillier Verlag, Göttingen, Germany.

Gawron, J.-C. and Theuvsen, L., 2009. The international food standard: bureaucratic burden or helpful management instrument in global markets? Empirical results from the German food industry. Journal of International Food and Agribusiness Marketing 21(4), 239-252.

Gawron, J.-C. and Theuvsen, L., 2010. Agrifood certification schemes in an intercultural context: theoretical reasoning and empirical findings. Food Economics – Acta Agriculturae Scandinavica, Section C 7(2), 209-221.

Gemünden, H.G., Ritter, T. and Heydebreck, P., 1996. Network configuration and innovation success: an empirical analysis in German high-tech industries. International Journal of Research in Marketing, 13, 449-462.

Gerboth, T., 2002. Statistische Prozessregelung bei administrativen Prozessen im Rahmen eines ganzheitlichen Prozesscontrollings. Dissertation. Technische Universität Berlin, Berlin, Germany.

Gilbert, M.J., Bos, M.E., Duim, B., Urlings, B.A., Heres, L., Wagenaar, J.A. and Heederik, D.J., 2012. Livestock-associated MRSA ST398 carriage in pig slaughterhouse workers related to quantitative environmental exposure. Occupational and Environmental Medicine 69(7), 472-478.

GIQS e.V., 2012. SAFEGUARD project news: Datenaustausch bringt Win-Win-Situation für Betriebe und Behörden. Available at: http://safeguard.giqs.org/aktuelles/aktuelles/datenaustausch-bringt-win-win-situation-fuer-betriebe-und-behoerden/.

Gold, S., Seuring, S. and Beske, P., 2010. Sustainable supply chain management and inter-organizational resources: a literature review. Corporate Social Responsibility and Environmental Management 17, 230-245.

Göser, T., 2010. Risikomanagement in Milchviehbetrieben und Konzeption eines Beratungsansatzes für die landwirtschaftliche Praxis. Master Thesis, Universität Göttingen, Göttingen, Germany.

Gramig, B.M., 2008. Essays on the economics of livestock disease management on-farm biosecurity adoption, asymmetric information in policy design, and decentralized bioeconomic dynamics. Dissertation. Michigan State University, East Lansing, MI, USA.

Green, P.E. and Srinivasan, V., 1978. Cojoint analysis in consumer research: issues and outlook. Journal of Consumers Research 5, 103-123.

Green, P.E., Krieger, A.M. and Agarwal, M.K., 1991. Adaptive conjoint analysis – some caveats and suggestions. Journal of Marketing Research 28(2), 215-222.

Griese, J. and Sieber, P., 1999. Betriebliche Geschäftsprozesse: Grundlagen, Beispiele, Konzepte. Haupt Verlag, Bern, Switzerland.

Grossekettler, H., 1978. Der Einfluss der innerorganisatorischen Informationsverarbeitung auf die Wachstumsstruktur einer Marktwirtschaft. Ein Test des Renditeausgleichstheorems und des Theorems vom konservativen Wachstum. In: Helmstädter, E. (ed.) Neuere Entwicklungen in den Wirtschaftswissenschaften. Schriften des Vereins für Sozialpolitik. Gesellschaft für Wirtschafts- und Sozialwissenschaften, Volume 98. Verlag Duncker and Humblot, Berlin, Germany, pp. 135-196.

Grüne, J., 1994. Anfänge staatlicher Lebensmittelüberwachung in Deutschland. Franz Steiner, Stuttgart, Germany.

GS1 Germany, 2012. GS1-Standards – Ein Lösungsportfolio weist in die Zukunft. GS1 Germany, Cologne, Germany.

Hansen, W. and Almeida, A., 1993. Zertifizierung und Akkreditierung von Produkten und Leistungen der Wirtschaft. Hanser Verlag, Munich, Germany, 158 pp.

Hardes, H.-D. and Uhly, A., 2007. Grundzüge der Volkswirtschaftslehre. 9th Edition, Oldenbourg Verlag, Munich, Germany.

Hartmann, H., 2004. Lieferantenmanagement – Gestaltungsfelder, Methoden, Instrumente mit Beispielen aus der Praxis. Deutscher Betriebswirte-Verlag, Gernsbach, Germany.

Hartmann, M., 2011. Corporate social responsibility in the food sector. European Review of Agricultural Economics 38(3), 297-324.

Heeschen, W. and Blüthgen, A., 1993. Evaluation of residues and contaminants in food: difficulties and significance. In: Sommer, H., Petersen, B. and Von Wittke, P. (eds.) Safeguarding food quality. Springer, Berlin, Germany, pp. 59-74.

Henry, J. and Walker, D. (eds.), 1991. Managing innovation. SAGE Publications, London, UK.

Herbert, U., Rossaint, S., Khanna, M.-A., Kreyenschmidt, J., 2013. Comparison of argon-based and nitrogen-based modified atmosphere packaging (MAP) on bacterial growth and product quality of chicken breast fillets. Poultry Science 92, 1348-1356.

Heres, L., Hoogenboom, R., Herbes, R., Traag, W. and Urlings, B., 2010. Tracing and analytical results of the dioxin contamination incident in 2008 originating from the Republic of Ireland. Food Additives and Contaminants Part A 27, 1733-1744.

Heyder, M., Fahrtmann, K. and Theuvsen, L.,2009. Lieferantenbewertung als Management-Instrument: Empirische Ergebnisse aus der Ernährungswirtschaft. In: Meyer, J.-A. (ed.) Management-Instrumente in kleinen und mittleren Unternehmen. Eul-Verlag, Lohmar, Germany, pp. 113-126.

Hiller, A., Heres, L., Schulze Althoff, G., Urlings, B. and Klein, G., 2012. Implementation of risk based meat inspection of fattening pigs without incision in a slaughterhouse in Lower Saxony. Archiv für Lebensmittelhygiene 63(4), 107-114.

Hiller, A., Oorburg, D., Wisselink, H.J., Van Solt-Smits, C.B., Urlings, B., Klein, G., Schulze Althoff, G. and Heres, L., 2013. Prevalence of *Mycobacterium avium* in slaughter pigs based on serological monitoring results. International Journal of Environmental Research and Public Health 10(9), 4027-4038.

Hirsch, B., 2010. Effzientes Management von Tierseuchenrisiken. Eine Analyse externer Effekte und asymmetrischer Information. Dissertation. Universität Hamburg, Hamburg, Germany.

Hoffmann, C., Breuer, O., Brinkmann, D., Schütz, V. and Petersen, B., 2008. Entwicklungstendenzen beim Aufbau privater und öffentlicher Frühwarnsysteme im überbetrieblichen Gesundheitsmanagement grenzübergreifender Zulieferketten. 28th GIL Annual Conference, Volume 125. Kiel, Germany, pp. 75-78. Available at: www.gil-net.de/Publikationen/20_75.pdf.

Hofstede, G.J. ,Spaans, L., Schepers, H., Trienekens, J.H. and Beulens, A.J.M. (eds.), 2004. Hide or confide: the dilemma of transparency. Reed BusinessInformation, Amsterdam, the Netherlands, 248 pp.

Hofstede, G.J., 2003a. Transparency in netchains. In: Harnos, Z., Herdon, M. and Wiwczaroski, T.B. (eds.) Information technology for a better agri-food sector, environment and rural living. Debrecen University, Debrecen, Hungary, pp. 17-29.

Hofstede, G.J., 2003b. Trust and transparency in supply netchains: a contradiction? Chapter 5. In: Wang, W.Y.C., Heng, M.S.H. and Chau, P.Y.K. (eds.) Supply chain management: issues in the new era of collaboration and competition. IGI Global, Hershey, PA, USA.

Hollmann-Hespos, T., 2008. Rückverfolgbarkeitssysteme in der Ernährungswirtschaft: Eine empirische Untersuchung des Investitionsverhaltens deutscher Unternehmen. Dr. Kovac, Hamburg, Germany.

Holm, V.K., Ndoni, S. and Risbo, J., 2006. The stability of poly(lactic) acid packaging films as influenced by humidity and temperature. Journal of Food Science 71(2), 40-44.

Hoornstra E. and Notermans, S., 2001. Quantitative microbiological risk assessment. International Journal of Food Microbiology 66, 21-29.

Horvath, L., 2004. Supply Chain Management in der Fleischerzeugung: Konzeption, Implementierung und Perspektive. Dissertation TU München. – Strategien, Methoden, Techniken. 3rd Edition, Carl Hanser Verlag, Munich, Germany.

Hotwagner, B., 2008. Supply Chain Risk Management und dessen systemische Umsetzung im Unternehmen. Wirtschaft und Management. In: Holzinger, H. (ed.) Wirtschaft und Management. Vienna, Austria, pp. 23-39.

Howells, J., 2006. Intermediation and the role of intermediaries in innovation. Research Policy 35, 715-728.

Hoy, S., Gauly, M., and Kiefer, J., 2006. Nutztierhaltung und -hygiene: Grundwissen Bachelor. UTB, Hohenheim, Germany.

Hummel, T and Malorny, C., 2002. Total Quality Management: Tipps für die Einführung. 3rd Edition, Carl Hanser Verlag, Munich, Germany.

Ibald, R., Dorau, C. and Bode, T., 2011. Mit Visionen dem Ziel näher kommen – Neue IT-Technologien zur inner- und überbetrieblichen Prozessoptimierung. Fleischwirtschaft(5), 51-58.

Ilg, Y. and Kreyenschmidt, J., 2011. Effects of food components on the antibacterial activity of polypropylene surfaces containing silver ions (Ag+). International Journal of Food Science and Technology 46, 1469-1476.

Ilg, Y., 2010. Antimicrobial methods for the use in contact with food. Südwestdeutscher Verlag für Hochschulschriften, Saarbrücken, Germany.

Ilg, Y., Kreyenschmidt, M.. Lorenz, R., Zerbe, J. and Kreyenschmidt, J., 2011. Assessment of sustainable antimicrobial polymers with regard to their applicability in the food chain. In: Taoukis, P.S., Stoforos, N.G., Karathanos, V.T. and Saravacos, G.D. (eds.) Proceedings of the 11th International Congress on Food Engineering and Food – Food Process Engineering in a Changing World, Volume I, Athens, Greece. May 22-26, pp. 99-100.

ISO 31000, 2009. Risk management – principles and guidelines. English version. Beuth Verlag, Berlin, Germany.

Jahn, G., Peupert, M. and Spiller, A., 2003. Einstellungen deutscher Landwirte zum QS-System: Ergebnisse einer ersten Sondierungsstudie. Discussion Paper 0302, Institute of Agricultural Economics, Universität Göttingen, Göttingen, Germany.

Jantke, C., 2006. Operative Risiken in Schlachtung und Fleischverarbeitung: Risikoanalyse und Konzept eines Risikomanagementsystems. Dissertation, Technische Universität Munich, Munich, Germany.

Jones, G.R. and George, J.M., 1998. The experience and evolution of trust: implications for cooperation and teamwork. Academy of Management Review 23, 531-546.

Kagerhuber, M. and Kühl, R., 2002. Unterschiede und Gemeinsamkeiten vertikaler Prozessorganisationen im Bereich der integrierten tierischen Produktion und Vermarktung. In: Landwirtschaftliche Rentenbank (ed.) Lebensmittelsicherheit und Produkthaftung. Frankfurt am Main, Germany, pp. 7-42.

Kamiske, G.F. and Brauer J.-P., 2008. Qualitätsmanagement von A bis Z. Erläuterungen moderner Begriffe des Qualitätsmanagements. Carl Hanser Verlag, Munich, Germany.

Kampa, H., 1996. Das Aus- und Weiterbildungskonzept für QM-Personal der DGQ. In: Aufbau und Zertifizierung von QM-Systemen in der Agrarwirtschaft – Lösungsbeispiele, Erfahrungen und Perspektiven, FCL Series, Volume 5.

Kampmann, Y., De Clerck, E., Kohn, S., Patchala, D., Langerock, R. and Kreyenschmidt, J., 2008. Study on the antimicrobial effect of silver-containing inner liners in refrigerators. Journal of Applied Microbiology 104, 1808-1814.

Karge, J., 2002. Untersuchung zur funktionellen Sicherheit von integrierten QM-Systemen in der Broiler-Produktion. Dissertation. Freie Universität Berlin, Berlin, Germany.

Kerry, J., O'Grady, M. and Hogan, S., 2006. Past, current and potential utilisation of active and intelligent packaging systems for meat and muscle-based products: a review. Meat Science 74, 113-130.

Kersten, W., Hohrath, P. and Winter, M., 2008. Risikomanagement in Wertschöpfungsnetzwerken – Status quo und aktuelle Herausforderungen. In: Wirtschaft und Management, Schriftenreihe zur wirtschaftswissenschaftlichen Forschung und Praxis, Volume 8, Supply Chain Risk Management, Fachhochschule des bfi Wien, Wien, Austria, pp. 7-21..

Killich, S., Bruns, I. and Künzer, A., 2007. In: Pfeiffer, T. and Schmitt, R. (eds.) Masing Handbuch Qualitätsmanagement, 5th edition. Carl Hanser Verlag, Munich, Germany, pp. 309-329.

Klauke, T.N., 2012. Risk based approach towards more sustainability in European pig production. Dissertation. Universität Bonn, Bonn, Germany.

Klauke, T.N. and Brinkmann, D., 2009. Kosteneffizienz durch Six Sigma Prozessoptimierung – Chancen für die Fleischproduktion. Fleischwirtschaft 8, 16-22.

Klerkx, L. and Leeuwis, C., 2008. Balancing multiple interests: embedding innovation intermediation in the agricultural knowledge infrastructure. Technovation 28, 364-378.

Klerkx, L., 2008. Matching demand and supply in the Dutch agricultural knowledge infrastructure. The emergence and embedding of new intermediaries in an agricultural innovation system in transition. Dissertation. Wageningen University, Wageningen, the Netherlands.

Köhler, P.T., 2006. Prince 2. Springer Verlag, Berlin, Germany.

Kreyenschmidt, J. and Ibald, R., 2012. Modeling shelf life using microbial indicators. In: Nicoli, M.C. (ed.) Shelf life assessment of food. CRC Press, Boca Raton, FL, USA, 127 pp.

Kreyenschmidt, J., 2003. Modellierung des Frischeverlustes von Fleisch sowie des Entfärbeprozesses von Temperatur-Zeit-Integratoren zur Festlegung von Anforderungsprofilen für die produktbegleitende Temperaturüberwachung. Dissertation. Universität Bonn, Bonn, Germany.

Kreyenschmidt, J., Albrecht, A., Braun, C., Herbert, U., Mack, M., Rossaint, S., Ritter, G., Teitscheid, P. and Ilg, Y., 2013. Food Waste in fleischerzeugenden Ketten. Fleischwirtschaft 10, 57-63.

Kreyenschmidt, J., Christiansen, H., Huebner, A, Raab, V. and Petersen, B., 2010. A novel photochromic time-temperature indicator to support cold chain management. International Journal of Food Science and Technology 45(2), 208-215.

Kreyenschmidt, J., Petersen, N., Petersen, B. and Kunz, B., 2003. Erprobung eines Temperatur-Zeit-Indikators im Rahmen von HACCP-Konzepten bei der Verpackung, Lagerung und dem Transport von Frischprodukten. Landwirtschaftliche Fakultät der Universität Bonn, Schriftenreihe des Lehr- und Forschungsschwerpunktes (USL), Vol. 106. Available at: http://www.usl.uni-bonn.de/pdf/Forschungsbericht%20106.pdf.

Lachenmeir, P., 2004. Chapter 5. In: Pfeufer, H.J., Schreiber, F. and Rau, W. (eds.) Das integrierte Managementsystem. Carl Hanser Verlag, Munich, Germany.

Ladeur, K.H. and Vesting, T., 2008. Geistiges Eigentum im Netzwerk – Anforderungen und Entwicklungslinien. In: Eifert, M. and W. Hoffmann-Riem (eds.) Geistiges Eigentum und Innovation. Duncker and Humblot, Berlin, Germany, pp. 123-144.

Lagerstedt, A., Ahnström, M.L. and Lundström, K., 2011. Vacuum skin pack of beef – a consumer friendly alternative. Meat Science 88, 391-396.

Lang, J., and Petersen, B., 2012. AMOR - Mutuale Prüfstrategien in komplexen Kunden- und Lieferantenbeziehungen der Agrar- und Ernährungswirtschaft . In: Woll, R. and Uhlemann, M. (eds.) Vielfalt Qualität - Tendenzen im Qualitätsmanagement. Shaker Verlag, Aachen, Germany, pp. 45-60.

Lazzarini, S.G., Chaddad, F.R. and Cook, M.L., 2001. Integration supply chains and network analysis, the study of net chains. Journal on Chain and Network Science 1, 7-21.

Lee, C.H., Park, H.J. and Lee, D.S., 2004. Influence of antimicrobial packaging on kinetics of spoilage microbiological milk and orange juice. Journal of Food Engineering 65, 527-531.

Lehnert, S., 1994a. Anstöße und Umsetzung der vertikalen Integration in Marken- und Qualitätsfleischerzeugung in Deutschland. In: Lehnert, S. (ed.) FCL-Schriftenreihe Volume 1 Controlling und Qualitätsmanagement in der Agrar- und Ernährungswirtschaft. Forschungsgemeinschaft Controlling in der Landwirtschaft e.V., Bonn, Germany, pp. 101-116.

Lehnert, S., 1994b. DIN ISO 9000ff zur Unterstützung der Eigeninitiative der Wirtschaft beim Aufbau von Qualitätsmanagementsystemen. In: Lehnert S. (ed.) FCL-Schriftenreihe Volume 1 Controlling und Qualitätsmanagement in der Agrar- und Ernährungswirtschaft. Forschungsgemeinschaft Controlling in der Landwirtschaft e.V., Bonn, Germany, pp. 67-80.

Lehnert, S., 1995a. AGRI-ZERT: Zertifizierungsstelle in der Agrarwirtschaft. In: Lehnert, S., Petersen, B., Jürgens, P. and Roux, N., (ed.) FCL-Schriftenreihe Volume 4 DIN EN ISO 9000ff.: Neue Impulse für das Qualitätsmanagement in der europäischen Agrarwirtschaft. Forschungsgemeinschaft Controlling in der Landwirtschaft e.V., Bonn, Germany, pp. 77-82.

Lehnert, S., 1995b. Vorbereitung landwirtschaftlicher Betriebe auf die Zertifizierung. In: Lehnert, S., Petersen, B., Kolb, B. und Jürgens, P. (ed.) FCL-Schriftenreihe Volume 3. Klein- und mittelständische Betriebe auf dem Weg zur Zertifizierung. Forschungsgemeinschaft Controlling in der Landwirtschaft e.V., Bonn, Germany, pp. 87-96.

Lehnert, S., 1996. QM-Systeme in der agrarwirtschaftlichen Kunden-Lieferanten-Beziehung. In: Jürgens, P., Petersen B. and Roux, N. (ed.) FCL-Schriftenreihe Volume 5 Aufbau und Zertifizierung von QM-Systemen in der Agrarwirtschaft Lösungsbeispiele, Erfahrungen und Perspektiven. Forschungsgemeinschaft Controlling in der Landwirtschaft e.V., Bonn, Germany, pp. 77-80.

Lehnert, S., 1998. Aufbau von Qualitätsmanagement-Systemen in landwirtschaftlichen Betrieben am Beispiel der Fleischproduktion. In: Lehnert, S. (ed.) FCL-Schriftenreihe Volume 6. Forschungsgemeinschaft Controlling in der Landwirtschaft e.V., Bonn, Germany, pp. 1-174

Lehnert, S., 2000. Verknüpfung von QM-Systemen in Produktionsketten. In: Lehnert, S. and Petersen, B., FCL-Schriftenreihe Volume 7. Anstoß und Vision – Qualitätsmanagement in der fleischerzeugenden Kette. Forschungsgemeinschaft Controlling in der Landwirtschaft e.V., Bonn, Germany, pp. 5-13.

Lehrner, J., 2002. Notwendigkeit, Nutzen und Realisierbarkeit eines Risiko-Managements in landwirtschaftlichen Betrieben. Dissertation. Universität Vienna, Vienna, Austria.

Leonard, K.-W. and Naumann, P., 2002. Managementsysteme – Begriffe. Deutsche Gesellschaft für Qualität e.V. (ed.) Vol.11-04. Berlin, Germany.

Lienemann, K. and Lehnert, S., 2005. Agrar- und Ernährungsforschung in Deutschland. Kooperationschancen erkennen – Innovationen nutzen. Ein Handbuch für Politik und Praxis: Band 2. Bergen/Dumme, Germany.

Lindgreen A, Palmer, R. and Trienekens, J.H., 2005. Relationships within a supply chain. Journal on Chain and Network Science 5(10), 85-101.

Lundvall, B.Å. (ed.), 1992. National systems of innovation. Towards a theory of innovation and interactive learning. Anthem Press, London, UK.

Luning, P.A. and Marcelis, W.J., 2009. Food quality management: technological and managerial principles and practices. Wageningen Academic Publisher, Wageningen, the Netherlands.

Mack, A., 2007. Nutzungskonzept für ein integriertes Audit- und Dokumentenmanagementsystem im überbetrieblichen Gesundheitsmanagement Schweine haltender Betriebe. Dissertation. Universität Bonn, Bonn, Germany.

Magnusson, K., Kroslid D. and Bergman, B., 2001. Six Sigma umsetzen. Die neue Qualitätsstrategie für Unternehmen. Carl Hanser Verlag, Munich, Germany.

Maloni, M.J. and Brown, M.E., 2006. Corporate social responsibility in the supply chain: an application in the food industry. Journal of Business Ethics 68, 35-52.

Malorny, C., 1997. Der Weg zum umfassenden Qualitätsmanagement. In: Spektrum der Wissenschaft(1), 96-99.

Manzini, R. and Accorsi, R., 2012. The new conceptual framework for food supply chain assessment. Journal of Food Engineering 115, 251-263.

Masing, W., 2007 Das Unternehmen im Wettbewerb. In: Pfeiffer, T. and Schmitt, R. (eds.) Masing Handbuch Qualitätsmanagement, 5th edition, Carl Hanser Verlag, Munich, Germany, pp. 3-15.

McMeekin, T.A., Bowman, J., McQuestin, O, Mellefont, L., Ross, T. and Tamplin, M, 2008. The future of predictive microbiology: strategic research, innovative applications and great expectations. International Journal of Food Microbiology 128(1), 2-9.

Meemken, D. and Blaha, T., 2008. Die Tiergesundheit im Fokus des neuen europäischen Lebensmittelsicherheitskonzeptes – Chancen für die Bestandsbetreuung durch den praktizierenden Tierarzt. Der praktische Tierarzt 89(1), 58- 63.

Menrad, K., 2004. Innovations in the food industry in Germany. Research Policy 33, 845-878.

Meuwissen, M.P.M., Velthuis, A.G.J., Hogeveen, H. and Huirne, R.B.M., 2003. Traceability and certification in meat supply chains. Journal of Agribusiness 21(2), 167-181.

Meyer, C., 2010. Qualitätskommunikation und Erfolgsfaktoren im Commodity-Marketing – Entwicklungsmöglichkeiten einer Getreidemarke mit regionaler Identität – Dissertation. Universität Bonn, Bonn, Germany.

Min, H. and Galle, W.P., 2001. Green purchasing practices of US firms. International Journal of Operations & Production Management 21, 1222-1238.

Mintzberg, H., 1979. The structuring of organisations: a synthesis of the research. Prentice Hall, Englewood Cliffs, NJ, USA.

Mousing, J., Jensen, P.T., Halgaard, C., Bager, F., Feld, N., Nielsen, B., Nielsen, J.P. and Bech-Nielsen, S., 1997. Nation-wide Salmonella enterica surveillance and control in Danish slaughter swine herds. Preventive Veterinary Medicine 29, 247-261.

Moynagh, J., 2000. EU regulation and consumer demand for animal welfare. AgBioForum 3(2-3), 107-114.

Nathues, H., Nienhoff, H., große Beilage, E., Blaha, T., Ritzmann, M.M., Reiner, G., Lahrmann, K.H., Kaufhold, J., Waberski, D., Hennig-Pauka, I., Wendt, M. and Waldmann, K.H., 2011. Monitoring-Systeme in Zuchtschweinebeständen aus Sicht der Wissenschaft. Deutsches Tierärzteblatt 10, 1324-1334.

National Annex to DIN EN ISO 9001, 2006. Interpretationen zu ISO 9001:2000 'Qualitätsmanagementsysteme — Anforderungen' vom 28. September 2006. Normenausschuss Qualitätsmanagement, Statistik und Zertifizierungsgrundlagen im DIN (NQSZ), Berlin, Germany. Available at: www.nqsz.din.de/sixcms_upload/media/2856/Interpretationen_ISO.35919.pdf.

Nienhoff, H.-J., 1995. Qualitätsmanagement auch für Landwirtschaft und Handwerk von Bedeutung. In: FCL-Schriftenreihe Volume 3, 'Klein- und mittelständische Betriebe auf dem Weg zur Zertifizierung', Landwirtschaftsverlag, Münster-Hiltrup, Germany.

Nienhoff, H.J., 2008. QS – Qualitätssicherung stufenübergreifend: Qualitätssicherung im Kontext der Marktanforderungen. In: Spiller, A. and Schulze, B. (eds.) Zukunftsperspektive der Fleischwirtschaft. Verbraucher, Märkte, Geschäftsbeziehungen. Universitätsverlag Göttingen, Göttingen, Germany, pp. 165-182.

NRLO – De Nationale Raad voor Landbouwkundig Onderzoek, 1999. Innoveren met ambitie, kansen voor de agrosector, groene ruimte en vissector. NRLO-rapport 99(17), NRLO, Den Haag, the Netherlands.

Nüssel, M., 1996. Erfahrungen eines Schweinemästers beim Aufbau eines QM-Systems und dessen Zertifizierung. In: FCL-Schriftenreihe Volume 5, 'Aufbau und Zertifizierung von QM-Systemen in der Agrarwirtschaft – Lösungsbeispiele, Erfahrungen und Perspektiven'. Landwirtschaftsverlag, Münster-Hiltrup, Germany.

Nychas, G.-J.E., Skandamis, P.N., Tassou, C.C. and Koutsoumanis, K.P., 2008. Meat spoilage during distribution. Meat Science 78(1-2), 77-89.

O'Hagan, J., Ellebrecht, S. and Petersen, B., 2013. Netchain coordinator services in pork supply chains including inspections according to the AMOR principle. Meat Science 95, 733-743.

ONR 49001, 2010. Risk management for organisations and systems – risk management – Practical use of ISO/DIS 31000. Austrian Standards Institute, Wien, Austria.

Opsteegh, M., Prickaerts, S., Frankena, K. and Evers, E.G., 2011. A quantitative microbial risk assessment for meatborne *Toxoplasma gondii* infection in the Netherlands. International Journal of Food Microbiology 150(2-3), 103-114.

Padel, S, Schaack, D. and Willer, L., 2009. Development of the organic market in Europe. In: Willer, H. and Kilcher, L. (eds.) The world of organic agriculture. Statistics and emerging trends. IFOAM - FiBL - ICT, Bonn - Frick - Geneva, Germany, Switzerland, pp. 155-163.

Padgett, T., Han, I.Y. and Dawson, P.L., 1998. Incorporation of food-grade antimicrobial compounds into biodegradable packaging Films. Journal of Food Protection 61, 1130-1335.

Petersen, B., 1985. Die Kontrolle von Leistung und Gesundheit in Ferkelerzeugerbetrieben mit Hilfe eines produktionsbegleitenden Informationssystems. Habilitation Universität Bonn. Verlag Orzekowsky, Bonn, Germany.

Petersen, B., 1995. Zielsetzung des Regelwerkes 9000 ff. Klein- und mittelständische Unternehmen auf dem Weg zur Zertifizierung. In: Lehnert, S., Petersen, B., Kolb, B. and Jürgens, P. (eds) Klein- und mittelständische Betriebe auf dem Weg zur Zertifizierung. FCL-Schriftenreihe, Volume 3. Forschungsgemeinschaft Controlling in der Landwirtschaft e.V., Bonn, Germany, pp. 1-11.

Petersen, B., 2003. Das Qualitätssicherungswesen in Deutschland und Europa – Rahmenbedingungen und Entwicklungen. In: Lebensmittelqualität und Qualitätssicherungssysteme. Agrarspectrum Schriftenreihe 37, pp. 35-48.

Petersen, b., 2012b. Qualitätskommunikation – Herausforderung und Chance zugleich! In: FoodNetCenter Bonn (ed.) Forschungsnetzwerk Innovation durch Qualitätskommunikation, FoodNetCenter, Bonn, Germany, pp. 18-25.

Petersen, B., Brinkmann, D., Slütter, S., Bruns, M., Kollmer, B., Festag, R., Münster, A. and Otto, M., 2010. Kommunikationskonzepte zwischen regionalen und bundesweiten Dienstleistungsagenturen. In: Petersen, B., Spiller, A. and Theuvsen, L. (eds.) Vom Viehvermarkter zum Dienstleistungsprofi. GIQS e.V., Bonn, Germany, pp. 287-298.

Petersen, B., Hartmann, M., Cremers, A. and Wüst, M. (eds.), 2012. FIN-Q.NRW – Forschungsnetzwerk Innovation durch Qualitätskommunikation. Projektdokumentation. Universität Bonn, International Center for Food Chain and Network Research (FoodNetCenter Bonn), Bonn, Germany. Available at: www.itw.unibonn.de/institut/pgm/forschung/projekte/fin-q.nrw-1/wissenschaftliche-broschuere/view.

Petersen, B., Knura-Deszczka, S., Pönsgen-Schmidt, E. and Gymnich, S., 2002. Computerised food safety monitoring in animal production. Livestock Production Science 76(3), 207-213.

Petersen, B., Lehnert, S. and Jürgens, P., 1997. Qualitätssicherung bei der Rohstofferfassung und Qualitätsmanagement im landwirtschaftlichen Betrieb. In: Einführung von Qualitätsmanagementsystemen nach ISO 9000 ff. in der landwirtschaftlichen Produktion und im Nahrungs- und Genussmittelgewerbe, FQS-DGQ Series 92-05, Beuth Verlag, Berlin, Germany, pp. 96-116.

Petersen, B., Mack, A., Schütz, V. and Schulze Althoff, G., 2007. Nahtstelle als neuralgischer Punkt – 3-Ebenen-Modell zur Weiterentwicklung überbetrieblicher Qualitätsmanagement-Systeme. Fleischwirtschaft(4), 89-94.

Pfeifer, T. and Schmitt, R. (eds.), 2007. Masing Handbuch Qualitätsmanagement. 5th completely revised edition, Carl Hanser Verlag, Munich, Germany.

Pfeifer, T., 1993. Qualitätsmanagement. Carl Hanser Verlag, Munich, Germany.

Pfeifer, T., 2001. Qualitätsmanagement: Strategien, Methoden, Techniken. 3rd Edition, Carl Hanser Verlag, Munich, Germany.

Pfeiffer, D.U., 2005. Can computerised information systems lead to more effective surveillance and monitoring of food safety? In: Smulders, F.J.M. and Collins, J.D. (ed.) Risk management strategies: monitoring and surveillance. Food Safety Assurance and Veterinary Public Health Series, Volume 3. Wageningen Academic Publishers, Wageningen, the Netherlands, pp. 69-80.

Phillips, C.A., 1996. Review: modified atmosphere packaging and its effects on the microbiological quality and safety of produce. International Journal of Food Science Technology 31, 463-479.

Pil, F.K. and Rothenberg, S., 2003. Environmental performance as a driver of superior quality. Production and Operations Management 12, 404-415.

Plumeyer, C.H., Deimel, M. and Theuvsen, L., 2008. Qualitätskommunikation und Prozessoptimierung in der Fleischwirtschaft: Recht, Zertifizierung und Informationssysteme als Einflussgröße. In: Goch, G. (ed.) Proceedings of GQW annual conference 2008. Shaker Verlag, Aachen, Germany, pp. 1-28.

Poignée, O., Hannus, T., Jahn, V. and Schiefer, G., 2005. Qm-G: Organisatorisch-technische Alternativen für Rückverfolgbarkeit und Qualitätssicherung in Unternehmen der Getreidewirtschaft. In: Schiefer, G. (ed.) Rückverfolgbarkeit und Qualitätsmanagement in der Getreide- und Futtermittelwirtschaft. Universität Bonn, Institut für landwirtschaftliche Betriebslehre (ILB), Bonn, Germany.

Poignee, O., Helbig, R. and Schiefer, G., 2002. Qualitätskommunikation in der Produktionskette: eine Fallstudie. In: Wild, K., Müller, R.A.E. and Birkner, U. (eds.) Proceedings of 23rd GIL annual conference. Gesellschaft für Informatik in der Land-, Forst- und Ernährungswirtschaft, Göttingen, Germany, pp. 171-174.

Pollard, D., 2006. Innovation and technology transfer intermediaries: a systemic international study. Advances in Interdisciplinary Studies of Work Teams 12, 137-174.

Porter, M.E., 1999. Wettbewerb und Strategie. Econ Ullstein List Verlag, Munich, Germany.

Quintavalla, S. and Vicini, L., 2002. Antimicrobial food packaging in meat industry. Meat Science 62, 373-380.

Raab, V., 2011. Assessment of novel temperature monitoring systems for improving cold chain management in meat supply chains. Dissertation. Universität Bonn, Bonn, Germany.

Raab, V., Bruckner, S., Beierle, E., Kampmann, Y., Petersen, B. and Kreyenschmidt, J., 2008. Generic model for the prediction of remaining shelf life in support of cold chain management in pork and poultry supply chains. Journal on Chain and Network Science 8(1), 59-73.

Raab, V., Ibald, R., Reichstein, W., Haarer, D., Petersen, B. and Kreyenschmidt, J., 2011. Novel solutions supporting inter-organisational quality and information management. In: Popov, V. and Brebbia, C.A. (eds.) 2011b. Food an environment – the quest of a sustainable future. Transactions on Ecology and the Environment 152, New Forest, WIT Press, Southhampton, UK, pp. 177-187.

Rammer, C., Zimmermann, V., Müller, E., Heger, D., Aschhoff, B. and Reize, F., 2006. Innovationspotenziale von kleinen und mittleren Unternehmen. ZEW Wirtschaftsanalysen, 79. Baden-Baden, Germany.

Reimer, A.P., Weinkauf, D.K. and Prokopy, L.S., 2012. The influence of perceptions of practice characteristics: An examination of agricultural best management practice adoption in two Indiana watersheds. Journal of Rural Studies 28(1), 118-128.

Rennings, K., 2000. Redefining innovation – eco-innovation research and the contribution from ecological economics. Ecological Economics 32, 319-332.

Riedel, U., 2009. Mitarbeiterqualifikation. In: Managementsysteme in der Lebensmittelwirtschaft, Loseblattsammlung (Version: 09 06 34). Behr's Verlag, Hamburg, Germany.

Rosenfeld, R. and Servo, J.C., 1991. Facilitating innovation in large organizations. In: Henry, J. and Walker, D. (eds.) Managing innovation. Sage, London, UK, pp. 28-39.

Rossaint, S. and Kreyenschmidt J., 2013. Assessment of TTI application to support cold chain management. In: Proceedings of the 5th International Workshop Cold Chain Management, 10-11 June 2013. Bonn, Germany. Available at: http://ccm.ytally.com/fileadmin/user_upload/downloads/publications_5th_workshop/Rossaint_paper.pdf.

Rothwell, R., 1994. Towards the fifth-generation innovation process. International Marketing Review 11(1), 7-31.

Rousis, K., Moustakas, K., Malamis, S., Papadopoulos, A. and Loizidou, M., 2008. Multicriteria analysis for the determination of the best WEEE management scenario in Cyprus. Waste Management 28(10), 1941-1954.

Rousseau, D., Sitkin, S., Burt, R. and Camerer, C., 1998. Not so different after all: a cross-discipline view of trust. Academy of Management Review 23, 393-404.

Ruiz-Capillas, C. and Jiménez-Colmenero, F., 2010. Effect of an argon-containing packaging atmosphere on the quality of fresh pork sausages during refrigerated storage. Food Control 21, 1331-1337.

Schaak, D., 2009. Five percent more organic land in the EU-27 - all crops on the increase. In: Willer, H. and Kilcher, L. (eds.). The world of organic agriculture. Statistics & emerging trends. IFOAM - FiBL - ICT, Bonn - Frick - Geneva, Switzerland, pp. 148-151.

Schenk, M., 1995. Soziale Netzwerke und Massenmedien: Untersuchungen zum Einfluss der persönlichen Kommunikation. Mohr Siebeck, Tübingen, Germany.

Schiefer, G. and Helbig, R., 1995. Qualitätsmanagement in der Agrarwirtschaft – Integration landwirtschaftlicher Betriebe in Entwicklungen zur Qualitätsproduktion in der Agrarwirtschaft. In: Neue Organisationsformen im Anpassungsprozeß der Landwirtschaft an die ökonomisch-technische Entwicklung in Produktion, Verarbeitung und Absatz. Schriftenreihe Landwirtschaftliche Rentenbank, Volume 9, Frankfurt am Main, Germany, pp. 69-112.

Schillings-Schmitz, A., 2006. HACCP und Lebensmittelhygiene – Antworten auf die häufigsten Fragen aus der Praxis. Behr's Verlag, Hamburg, Germany.

Schlicksupp, H., 1992. Innovation, Kreativität und Ideenfindung. Vogel Business Media, Würzburg, Germany.

Schmelzer, H.J. and Sesselmann, W., 2008. Geschäftsprozessmanagement in der Praxis – Kunden zufriedenstellen, Produktivität steigern, Wert erhöhen. Carl Hanser Verlag, Munich, Germany.

Schmitt, R., Betzold, M. and Hense K. 2007. In: Pfeiffer, T. and Schmitt, R. (eds.) Masing Handbuch Qualitätsmanagement, 5th edition. Carl Hanser Verlag, Munich, Germany, pp. 38-41.

Schmitz, T., 2006. Entwicklung eines Modells zum Einsatz präventiver Qualitätsmanagementmethoden in der Beratung von Unternehmen der Agrar- und Ernährungswirtschaft. Dissertation. Universität Bonn, Bonn, Germany.

Schouwenburg, H., 1995. Durchführung des IKB-Konzeptes in den Niederlanden vom Pilotprojekt bis zur Qualitätsstrategie. In: FCL-Schriftenreihe Volume 4 'DIN EN ISO 9000ff.: Neue Impulse für das Qualitätsmanagement in der europäischen Agrarwirtschaft'. Landwirtschaftsverlag, Münster-Hiltrup, Germany.

Schramm, M. and Spiller, A., 2003. Farm-Audit und Farm-Advisory-System – Ein Beitrag zur Ökonomie von Qualitätssicherungssystemen. Berichte über Landwirtschaft 81(2), 165-191.

Schroeder, R., Van de Ven, A., Scudder, G. and Polley, D., 1986. Managing innovation and change processes: findings from the Minnesota innovation research program. Agribusiness 2(4), 501-523.

Schulte-Wülwer. J. (ed.), 2012. Leitfaden zum Monitoring bei Ferkeln in den Niederlanden und in Deutschland. GIQS e.V., Bonn, Germany. Available at: http://safeguard.giqs.org/fileadmin/web_safeguard/content/Berichte_Protokolle2009/Schlussdokumente/Ringtest_dt_kl.pdf.

Schulze Althoff, G., 2004. Informationsmanagement in Netzwerken und Ketten der Schweinefleischerzeugung – Herausforderung, Möglichkeiten, Grenzen. In: Herrmann J. (ed) Qualitätsmanagement 6, Shaker Verlag, Aachen, Germany.

Schulze Althoff, G., 2006. Stufenkonzept zum Aufbau überbetrieblicher Informationssysteme für das Qualitäts- und Gesundheitsmanagement in Wertschöpfungsketten der Fleischwirtschaft. Dissertation. Universität Bonn, Bonn, Germany.

Schulze Althoff, G., Breuer, O., Jaeger, F. and Petersen, B., 2006. Working together to improve early warning. In: Fleischwirtschaft international, 2, Reprint: Deutsches Tierärzteblatt 11, 1332-1337.

Schulze Althoff, G., Ellebrecht, A. and Petersen, B., 2005. Chain quality information management - development of a reference model for quality information requirements in pork chains. Journal on Chain and Network Science 5(1), 27-38.

Schulze Althoff, G., Schulpin, C., Schmitz-Eiberger, M. and Petersen, B., 2007. Einführung von dynamischen Minimierungsansätzen für Rückstände in Frischeprodukten. Journal für Verbraucherschutz und Lebensmittelsicherheit 2(1), 20-33.

Schulze, H., Albersmeier, F., Gawron, J.-C., Spiller, A. and Theuvsen, L., 2008. heterogeneity in the evaluation of quality assurance schemes: the international food standard (IFS) in European agribusiness. International Food and Agribusiness Management Review 11(3), 99-139.

Schulze, H., Albersmeier, F., Spiller, A. and Jahn, G., 2007. Checklist Governance? Zur Prüfungsqualität von Zertifizierungssystemen im Agribusiness. In: Kuhlmann, F. and Schmitz, M. (eds.) Good Governance in der Agrar- und Ernährungswirtschaft. Landwirtschaftsverlag, Münster-Hiltrup, Germany, pp. 215-225.

Schütz, V. and Petersen, B., 2009. Planning model supporting information and service agency in meat chains designing services in chain oriented health management. Proceedings of the 19th IAMA Congress, 19-20 June 2009. European Organization for Quality, Budapest, Hungary. Available at: www.eoq.hu/iama/conf/1204_paper.pdf.

Schütz, V. and Petersen, B., 2010. Koordinationsaufgaben im überbetrieblichen Gesundheitsmanagement. Vom Viehvermarkter zum Dienstleistungsprofi, GIQS e.V., Bonn, Germany, pp. 163-182.

Schütz, V., 2009. Modell zur Planung von Dienstleistungen für das überbetriebliche Gesundheitsmanagement in der Fleischwirtschaft. Dissertation. Universität Bonn, Bonn, Germany.

Schweer, M. and Thies, B., 2003. Vertrauen als Organisationsprinzip. Hans Huber, Bern, Switzerland.

Schwerin, M., Balmann, A., Baum, M., Born, H., Mettenleiter, T. C., Patermann, C., Preisinger, R., Rodehutscord, M., Schulz, C., Swalve, H. and Taube, F., 2010. Herausforderungen für eine zukunftsfähige Erzeugung von Lebensmitteln tierischer Herkunft. Positionspapier der Arbeitsgruppe Tier des BioÖkonomieRats, Schriftenreihe Berichte aus dem BioÖkonomieRat 3, BioÖkonomieRat, Berlin, Germany.

Shapiro, D., Sheppard, B.H. and Cheraskin, L., 1992. Business on a handshake. Negotiation Journal 8, 365-377.

Sharfman, M.P., Shaft, T.M. and Anex Jr., R.P., 2009. The road to cooperative supply-chain environmental management: trust and uncertainty among pro-active firms. Business Strategy and the Environment 18, 1-13.

Simon, P., Chaudhry, Q. and Bakos, D., 2008. Migration of engineered nanoparticles from polymer packaging to food – a physiochemical view. Journal of Food and Nutritional Research 47, 105-113.

Slütter, S., 2013. Approach to use the engage exchange model for information management in crisis communication and animal disease control. Dissertation. Universität Bonn, Bonn, Germany. Available at: http://hss.ulb.uni-bonn.de/2013/3278/3278.htm.

Slütter, S., Petersen, B., Schütz, V. and Brinkmann, D., 2010. Reifegrad von Informations- und Kommunikationsdienstleistungen im überbetrieblichen Gesundheitsmanagement. In: Petersen, B., Spiller, A. and Theuvsen, L. (eds.) Vom Viehvermarkter zum Dienstleistungsprofi. GIQS e.V., Bonn, Germany, pp. 213-220.

Smid, J.H., Heres, L., Havelaar, A.H. and Pielaat, A.A., 2012. Biotracing model of *Salmonella* in the pork production chain. Journal of Food Protection 75(2), 270-80.

Smith, D. and Sparks, L., 2004. Temperature controlled supply chains food manufacturers can ill-afford inaccurate extrapolations based on a limited knowledge base. In: Bourlakis, M.A. and Weightman, P.W.H. (eds.) Food supply chain management. Blackwell Publishing, Oxford, UK, pp. 179-198.

Spiller, A., 2010. Gebündeltes Kundenbeziehungsmanagement und Logistik. In: Petersen, B., Spiller, A. and Theuvsen, L. (eds.) Vom Viehvermarkter zum Dienstleistungsprofi. GIQS e.V., Bonn, Germany, p. 91-160.

Spiller, A., Theuvsen, L. Recke, G. and Schulze, B., 2005. Sicherstellung der Wertschöpfung in der Schweineerzeugung: Perspektiven des Nordwestdeutschen Modells. Stiftung Westfälische Landschaft, Münster, Germany.

Stock, J.R. and Lambert, D.M., 2001. Strategic logistics management 4, McGraw-Hill/Irwin, Boston, MA, USA.

Strotmann, K., 1989. Modell zur Verarbeitung von Ergebnissen aus Gesundheitsvorsorge und Trächtigkeitstests zu vorhersagenden Informationen als Entscheidungshilfen für den Sauenhalter. Dissertation. Universität Bonn, Bonn, Germany.

Swanenburg, M., Van der Wolf, P.J., Urlings, H.A., Snijders, J.M. and Van Knapen, F., 2001. Salmonella in slaughter pigs: the effect of logistic slaughter procedures of pigs on the prevalence of *Salmonella* in pork. International Journal of Food Microbiology 70, 231-242.

Taoukis, P., 2008. Application of time-temperature integrators for monitoring and management of perishable products. In: Kerry, J. and Butler, P. (eds.) Smart packaging technologies for fast moving consumer goods. John Wiley and Sons, West Sussex, UK, 61 p.

Taoukis, P.S., Koutsoumanis, K. and Nychas, G.J.E., 1999. Use of time-temperature integrators and predictive modelling for shelf life control of chilled fish under dynamic storage conditions. International Journal of Food Microbiology 53(1), 21-31.

Taylor, A.A., Down, N.F. and Shaw, B.G., 1990. A comparison of modified atmosphere and vacuum skin packing for the storage of red meats. International Journal of Food Science Technology 25, 98-109.

The Institute of Risk Management (IRM), 2002. A risk management standard. The Institute of Risk Management (IRM), The National Forum for Risk Management in the Public Sector (ALARM), The Association of Insurance and Risk Managers (AIRMIC). Available at: http://www.theirm.org/publications/documents/Risk_Management_Standard_030820.pdf.

Theuvsen, L. and Friedrich, N., 2012. Vom Qualitäts- zum Nachhaltigkeitsmanagement: Wo steht das deutsche Agribusiness? In: Woll, R. and Uhlemann, M. (eds.) Vielfalt Qualität – Tendenzen im Qualitätsmanagement. Shaker Verlag, Aachen, Germany, pp. 319-338.

Theuvsen, L., 2003. Rückverfolgbarkeit von Lebensmitteln: Herausforderungen und Lösungsansätze aus organisatorischer Sicht. Berichte über Landwirtschaft 81, 555-581.

Theuvsen, L., 2004. Transparency in netchains as an organisational phenomenon: exploring the role of interdependencies. Journal on Chain and Network Science 4, 125-138.

Theuvsen, L., Spiller, A., Peupert, M. and Jahn, G. (eds.), 2007. Quality management in food chains. Wageningen Academic Publishers, Wageningen, the Netherlands.

Thomas, P.R., 1990. Competitiveness through total cycle time: an overview for CEOs. McGraw-Hill, New York, NY, USA.

TiGA (ed.), 2012. TiGA-Standard Leitfaden Stand 01.04.2012. Available at: https://www.tiergesundheitsagentur.de/downloads/TiGA/TiGA_Standard.pdf.

Tomankova, J., Borilova, G., Steinhauserova, I. and Las, L.G., 2012. Volatile organic compounds as biomarkers of the freshness of poultry meat packaged in a modified atmosphere. Czech Journal of Food Science 30, 395-403.

Trienekens J. and Beulens A., 2001a. The implication of EU food safety legislation and consumer demands on supply chain information systems. IFAMA, Washington, DC, USA. Available at: http://ifama.org/events/conferences/2001/cmsdocs/Trienekens_Jacques.pdf.

Trienekens J.H. and Van der Vorst, J. 2005. Traceability in food supply chains. In: Luning, P.A., DeVlieghere, F. and Verhe, R. (eds) Safety in the agri-food chain. Wageningen Academic Publishers, Wageningen, the Netherlands, pp. 439-471.

Trienekens, J., Petersen, B., Wognum, N. and Brinkmann, D. (eds.), 2009. European pork chains: diversity and quality challenges in consumer-oriented production and distribution. Wageningen Academic Publishers, Wageningen, the Netherlands.

Trienekens, J.H. and Beulens, A.J.M., 2001b. Views on inter-enterprise relationships. Production Planning & Control 12(5), 466-477.

Trienekens, J.H., 1999. Management of processes in chains: a research framework. PhD thesis. Wageningen University, Wageningen, the Netherlands.

Trommsdorff, V., 1990. Innovationsmanagement in kleinen und mittleren Unternehmen. Grundzüge und Fälle – Ein Arbeitsergebnis des Modellversuchs Innovationsmanagements. Munich, Germany.

United Nations Environment Programme (UNEP), 2012. Annual Report 2012. UNEP, Nairobi, Kenya.

United States Department of Agriculture (USDA), 2001. Biosecurity on dairies. Animal Plant Health Inspection Service (APHIS) Web site. Available at: http://www.aphis.usda.gov/animal_health/nahms/dairy/downloads/bamn/BAMN01_BiosecurityDairies.pdf.

Vachon, S. and Klassen, R.D., 2008. Environmental management and manufacturing performance: the role of collaboration in the supply chain. International Journal of Production Economics 111, 299-315.

Vahs, D. and Burmester, R., 2005. Innovationsmanagement. Von der Produktidee zur erfolgreichen Vermarktung. Schäffer-Pöschel Verlag, Stuttgart, Germany.

Van Dijk, S.J., Duysters, G.M. and Beulens, A.J.M., 2003. Transparency dilemmas in strategic alliances. Working Paper, KLICT, s'Hertogenbosch, the Netherlands.

Van Doorn, J. and Verhoef, P.C., 2011. Willingness to pay for organic products: differences between virtue and vice foods. International Journal of Research in Marketing 28(3), 167-180.

Van Hoek, A.H., De Jonge, R., Van Overbeek, W.M., Bouw, E., Pielaat, A., Smid, J.H., Malorny, B., Junker, E., Löfström, C., Pedersen, K., Aarts, H.J. and Heres, L., 2012. A quantitative approach towards a better understanding of the dynamics of Salmonella spp. in a pork slaughter-line. International Journal of Food Microbiology 153(1), 45-52.

Van Wagenberg, C.P.A., 2010. Incentive mechanisms for food safety control in pork supply chains. Dissertation. Wageningen University, Wageningen, the Netherlands.

Vanini, U., 2012. Risikomanagement. Schäffer-Poeschel Verlag, Stuttgart, Germany.

Vázquez, B.I., Carriera, L., Franco, C., Fente, C., Cepeda, A. and Barros-Velázquez, J., 2004. Shelf life extension of beef retail cuts subjected to an advanced vacuum skin packaging system. European Food Research Technology 218, 118-122.

Vermeiren, L., Devlieghere, F., Van Beest, M., De Kruijf, N. and Debevere, J., 1999. Developments in the active packaging of foods. Trends in Food Science and Technology 10, 77-86.

Von Borell, E., Bockisch, F.-J., Büscher, W., Hoy, S., Krieter, J., Müller, C., Parvizi, T., Richter, T., Rudovsky, A., Sundrum, A. and Van den Weghe, H., 2001. Critical control points for on-farm assessment of pig housing. Livestock Production Science 72, 177-184.

Wagner, K.W. and Dürr, W., 2008. Reifegrad nach ISO/IEC 15504 (SPICE) ermitteln. Carl Hanser Verlag, Munich, Germany.

Wappis, J. and Jung B., 2008. Taschenbuch Null-Fehler-Management. Carl Hanser Verlag, Munich, Germany, pp. 1-12.

Weber, M., 1922. Wirtschaft und Gesellschaft: Grundriss der verstehenden Soziologie. Mohr Siebeck, Tübingen, Germany.

Wellems, E., 2009. Pro & Contra: Einsatz externer Qualitätsmanagementbeauftragter – nützlich oder unsinnig? QZ 54(6), 14-17.

Wells, J.H. and Singh, R.P., 1989. A quality-based inventory issue policy for perishable foods. Journal of Food Processing and Preservation 12(4), 271-292.

Wells, S.J., 2000. Biosecurity on dairy operations: hazards and risks. Journal of Dairy Science 83(10), 2380-2386.

Wilke, T., Belaya, V. and Pinior, B., 2012a. How to measure food safety? A review of relevant literature. In: Rickert, U. and Schiefer, G. (eds.) Proceedings in System Dynamics and Innovation in Food Networks 2012. International European Forum (Igls-Forum) on System Dynamics and Innovation in Food Networks. Innsbruck-Igls, Austria, February 13-17, 2012, p. 61-83. Available at: http://centmapress.ilb.uni-bonn.de/ojs/index.php/proceedings/article/view/223/202.

Wilke, T., Düsseldorf, S., and Petersen, B., 2012b. Factors influencing on-farm-biosecurity measures: steps towards on-farm-biosecurity management. In: Stegeman, A. and Dewulf, J. (eds.) Book of Abstracts of the 13th International Symposium on Veterinary Epidemiology and Economics (ISVEE), Wageningen Academic Publishers, Wageningen, the Netherlands, p. 198.

Wilke, T., Raab, V., Breuer, O., Hamer, M. and Petersen, B., 2013. Lebensmittelsicherheit und die Entwicklung von Public-Private-Partnership-Strukturen. Zeitschrift für Außen- und Sicherheitspolitik 6, 23-33.

Willer, H., 2009. Organic agriculture in Europe: overview. In: H. Willer and Kilcher, L. (eds.) The world of organic agriculture. statistics & emerging trends 2009. IFOAM - FiBL - ICT, Bonn - Frick - Geneva, Switzerland, pp. 152-154.

Williamson, O.E., 1985. The economic institutions of capitalism. The Free Press, New York, NY, USA.

Winch, G.M. and Courtney, R., 2007. The organization of innovation brokers: an international review. Technology Analysis & Strategic Management 19(6), 747-763.

Wisselink, H.J., Van Solt-Smits, C.B., Oorburg, D., Van Soolingen, D., Overduin, P., Maneschijn-Bonsing, J., Stockhofe-Zurwieden, N., Buys-Bergen, H., Engel, B., Urlings, B.A. and Thole, J.E., 2010. Serodiagnosis of *Mycobacterium avium* infections in pigs. Veterinary Microbiology 142(3-4), 401-407.

Wissenschaftliche Beiräte für Verbraucher und Ernährungspolitik sowie Agrarpolitik des BMELV, 2011. Politikstrategie Food Labelling. September 2011. BMELV, Berlin, Germany.

Wittink, D.R., 1994. Commercial use of conjoint analysis in Europe: results and critical reflections. International Journal of Research in Marketing 11(1), 41-52.

Wocken, C. and Spiller, A., 2009. Gestaltung von Milchlieferverträgen: Strategien für die Molkereiwirtschaft nach Auslaufen der Quote. In: Theuvsen, L. and Schaper, C. (eds.) Milchwirtschaft ohne Quote. Märkte und Strategien im Wandel. Eul Verlag, Lohmar, Germany, pp. 111-132.

Wognum, P.M., Bremmers, H., Trienekens, J.H., Van der Vorst, G.A.J. and Bloemhof, J.M., 2011. Systems for sustainability and transparency of food supply chains – current status and challenges. Advanced Engineering Informatics 25, 65-76.

Wohlgemuth, O., 2002. Management netzwerkartiger Kooperationen – Instrumente für die unternehmensübergreifende Steuerung. Deutscher Universitäts-Verlag, Wiesbaden, Germany.

Woidich, A., 2004. Praxishandbuch Lebensmittelsicherheit. Behr's Verlag, Hamburg, Germany.

World Commission on Environment and Development (WCED), 1987. Our common future. WCED, New York, NY, USA.

Zeithaml, V.A., Parasuraman, A. and Berry, L.L, 1990. Delivering quality service. The Free Press, New York, NY, USA.

Zhu, Q., Geng, Y. and Lai, K., 2010. Circular economy practices among Chinese manufacturers varying in environmental-oriented supply chain cooperation and the performance implications. Journal of Environmental Management 91, 1324-1331.

Zhu, Q., Sarkis, J. and Geng, Y., 2005. Green supply chain management in China: pressures, practices and performance. International Journal of Operations & Production Management 25, 449-468.

Zhu, Q., Sarkis, J. and Lai, K., 2008b. Green supply chain management implications for „closing the loop". Transportation Research Part E: Logistics and Transportation Review 44, 1-18.

Zhu, Q., Sarkis, J. and Lai, K.-H., 2008a. Confirmation of a measurement model for green supply chain management practices implementation. International Journal of Production Economics 111, 261-273.

Zucker, L.G., 1986. Production of trust: institutional sources of economic structure, 1840-1920. Research in Organizational Behavior 8, 53-111.

Keyword index

B. Petersen, M. Nüssel and M. Hamer (eds.) *Quality and risk management in agri-food chains*
DOI 10.3920/978-90-8686-789-9_10, © Wageningen Academic Publishers 2014

N

O

P

T

U

V

W